# Design History

## HERMES

*Hermes* began life as a 1915 design study for a seaplane carrier capable of operating with cruisers in the reconnaissance role. This concept was changed in 1917 when the Admiralty created requirements for a purpose-built aircraft carrier with a 500ft flight deck, an internal hangar with capacity for 28 aircraft, islands to port and starboard connected by a navigating bridge, speed of 26 knots, two lifts, protection against cruiser gunfire, ten 6in guns for defence, four 4in AA guns, the ability to handle seaplanes over the stern, and a rotating catapult or rotatable forward portion of the flight deck to allow aircraft launches without having to turn the ship into the wind.

The catapult and rotatable flight deck were dropped from this extensive shopping list in favour of a 90ft wide flight deck that would allow athwart-ship takeoffs. Detailed design drawings were completed in January 1918 and the ship was laid down at Armstrong's Elswick yard on the 17th. Construction proceeded in fits and starts but, completed to the hangar deck level, the ship was launched on 11 September 1919 and towed to Devonport in January 1920 as the Elswick yard was closing.

The ship was laid up for two years while the design was reworked as a result of the successful trials of HMS *Eagle*. A single starboard side island was approved in March 1920, 5.5in guns were substituted for the 6in, and the armour protection scheme was dropped. A proposal to extend the hangar out to the sides of the ship in order to accommodate more aircraft was turned down on the grounds that it was structurally impossible given the basic design. Building resumed in April 1922, and the ship was commissioned in February 1924, but her protracted construction meant that while she was the first ship in the world to be designed and launched as an aircraft carrier, the Japanese *Hôshô* was the first ship designed as an aircraft carrier to enter service, which was in 1922.

The ship's most distinguishing feature was the massive island and heavy tripod mast which supported a gunnery control top, directors for the 5.5in guns, and a High Angle control position. The island housed

*Hermes* dressed overall at the July 1924 Fleet Review. The longitudinal arrester gear can be plainly seen. The aircraft on the flight deck are Fairey Flycatchers and Parnall Panthers; the Panthers have the open slot in the upper wing. *(Seaforth collection)*

ship navigating positions, sea cabins, searchlights, flying control, and communications offices; the bridge was fitted with a bullet proof shelter. The crane aft of the island was rated at 7,850lbs and could not lift a fully-fuelled float equipped Swordfish.

Constructed of 1in high tensile steel, the flight deck was the main strength deck and measured 570ft long by 90ft wide, tapering to a point at the bow and with a pronounced round down aft. The distinctive hump just forward of the after lift well was the attachment point for her original set of fore and aft arrester wires. The hump was left in place when that gear was removed in 1928.

The 400ft long hangar was fully enclosed within the hull by bulkheads at the front, rear, and along the sides; the rear bulkhead was fitted with a roll-up door that gave access to the quarterdeck. There were also three metal fire-proof curtains that rolled down from the deckhead. Passageways ran fore and aft outboard of the hangar sides, providing space for the 5.5in guns and boat stowage. Two lifts were fitted (46ft x 47ft), both cruciform shaped to accommodate aircraft with non-folding wings. The after lift lowered to the quarterdeck where the hangar was accessible through the roll-up door. Seaplanes could be embarked over the quarterdeck and moved directly into the hangar or hoisted to the flight deck on the after lift.

7000 gallons of aircraft petrol was carried in two cylindrical tanks on the hold deck with fuel delivered by a water pressure system. An additional 500 gallons was stored in a ready use tank on the starboard side of the island which could be jettisoned in an emergency.

To improve seakeeping the forward section of the hull was plated in, flaring considerably from waterline to flight deck in order to keep the flight deck as wide as possible. Protection was limited to a 2in belt of high tensile steel at the waterline. Bulges were fitted port and starboard.

The defensive armament consisted of six single 5.5in low angle mounts, fitted three to a side in apertures just below the flight deck. Four single 4in high angle mounts made up her entire anti-aircraft armament, fitted on the starboard side of the flight deck with two forward and two aft of the island.

As completed *Hermes* had a standard displacement of 10,850 tons with an overall length of 600ft and achieved 26.178kts at 41,318shp on trials. Endurance was 4000nm at 18kts. She proved to be successful in service and was a very good sea boat, despite the ungainly appearance of the island and mast.

## ARK ROYAL

Just the second British carrier to be designed and built from the keel up, *Ark Royal* was the culmination of all that had been learned from the development and operation of the preceding generation of carriers.

Detailed design work began in 1931. The requirements were many: an 800ft main flight deck, double flight decks as in HMS *Glorious*, able to handle seaplanes on the quarterdeck, stowage for 72 aircraft, an island, 32 knots top speed, two large lifts,

*Hermes* at anchor in Cawsand Bay in September 1933. The ship is in her China Station colours; note the island petrol tank and the swung-out navigating bridge. *(© National Maritime Museum, Greenwich, London N10525)*

protection over the vitals from 6in shell and 500lb bombs, from 20lb bombs at the flight deck, and from underwater charges of 750lb, a heavy anti-aircraft battery, endurance of 10,000nm at 10 knots, and 100,000 gallons of aviation fuel; to be achieved on a standard displacement of 22,000 tons.

The standard displacement was an arbitrary figure that the British government was hoping to get the other Naval Treaty powers to agree with in 1932. In the event, no agreement was reached but the limit remained in place for the new carrier.

The speed requirement was dropped to 30 knots and the double flight deck was abandoned in order to get everything to fit within the chosen displacement. Due to fiscal constraints, the ship was not ordered until 1935 which gave time for further design refinements including cranes fitted amidships for hoisting seaplanes. This eliminated the use of the quarterdeck for flight operations and enabled the stern to be fully plated in up to the flight deck.

The ship was laid down at the Cammell Laird yard in Birkenhead on 16 September 1935, launched on 13 April 1937, and commissioned on 16 November 1938. With her high freeboard, island, and fully plated in hull, *Ark Royal* was a radical departure from previous British carriers. The island was aerodynamically designed for improved air flow. It had four deck levels and contained the wheelhouse, captain's sea cabin, air office, navigating bridge and chart office, enclosed admiral's bridge, and signal bridge.

The Type 72X homing aerial in its characteristic dustbin-shaped wooden cover was atop the tripod mast, the cover preventing direct observation of the aerial which would enable its transmitting wavelength to be worked out. The aerial rotated at 1rpm and sent out a pulse every second. It was very sensitive to electrical interference which is why *Ark Royal* was never fitted with radar.

The flight deck extended 75ft beyond the stern in order to fulfil the requirement for an 800ft deck. Constructed of ¾in D quality steel amidships and thinning to ⅝in at the ends, it was proof against 20lb bombs. The round downs fore and aft used 80ft of deck space that aircraft could not be parked on.

Two BH3 catapults were fitted forward, each capable of accelerating 11,000lbs to 66 knots. Aircraft were launched tail up atop a trolley attached to a shuttle. The shuttle was accelerated the length of the catapult track via the action of compressed air on a piston within a sealed cylinder, the piston being attached to the shuttle via a set of wires and pulleys to gain an 8-1 mechanical advantage.

Eight transverse arrester wires were fitted aft, each capable of stopping 11,000lbs from 55 knots within 140ft. A safety barrier was fitted forward of the arrester wires, the first on a British carrier.

A 1939 view of *Ark Royal* with Swordfish aft and Ospreys forward shows how much of the after flight deck could not be used for spotting strikes. The safety netting around the after round down is very apparent. *(Seaforth collection)*

Eight antenna masts were fitted at the edges of the flight deck.

To obtain the space needed to meet the requirements for aircraft stowage, two hangars were built into the ship: the upper was 568ft long, the lower 452ft. Both were 60ft wide and 16ft high. Each was fully enclosed with its own ventilation and drainage systems, with four roll-down metal fire-proof curtains in the upper and three in the lower. The large hull openings did not extend into either hangar, but into compartments, passageways, and boat stowage galleries outboard of the hangar side bulkheads.

Three lifts were fitted inside the hangar against the outer bulkheads; two on the starboard side and one to port with the port side lift situated between the other two. The forward lift was 45ft x 25ft, the other two being 45ft x 22ft; all could lift 14,000lbs. Each lift had two independent platforms with each platform serving two decks only; the upper platform served the flight deck and the upper hangar, the lower served the upper and lower hangars. This allowed a short cycle time of 20 seconds from one deck to another, but proved awkward when moving aircraft from the lower hangar to the flight deck. Roll-down fire curtains were fitted on both hangar decks around the lift wells.

*Ark Royal* newly completed at Portsmouth in 1939. Many features are clearly shown including the hydraulic rams for the side windbreak, the crane folded against the hull above the first set of hull openings, stowed boat booms, the shield between S1 and S2 pompoms, safety netting, two High Angle Directors, and the bridge wind deflector. *(© National Maritime Museum, Greenwich, London N10552)*

The side protection was a 4.5in armour belt amidships 312ft long with its top edge mostly at the upper deck level (the deck below the lower hangar deck). It stepped up to the lower hangar deck abreast the island in order to cover the tops of the boilers and a portion of the funnel uptakes.

Across the width of the ship atop the side belt was a 3.5in armoured deck, stepping up to the lower hangar deck abreast the island with the side belt, then dropping to the lower deck directly over the magazines forward of the island. At each end of the side belt, a 3in armoured transverse bulkhead ran across the width of the ship; the whole scheme created an armoured box around the magazines, machinery spaces, and the aviation fuel tanks. A separate armoured box protected the main steering gear aft.

The torpedo protection was an air-liquid-air 'sandwich' of compartments along the ship's side from the waterline down to the turn of the bilge. Viewed in cross section, the outermost and innermost compartments were empty, with liquid filled compartments between them. A 1.5in anti-splinter bulkhead formed the innermost bulkhead. The liquid filled compartments contained fuel oil which could be replaced by water as it was used.

The layout of the machinery plant was complicated by the requirement for high speed, a desire for the shortest possible armoured box in order to save weight, and a maximum hull length of 725ft in order to fit within as many existing drydocks as possible. The result was a three-shaft arrangement; two shafts could not provide enough power and there was not enough room for four shafts. 102,000shp was provided by six boilers grouped in pairs in three boiler rooms; the propeller shafts were driven by three turbine sets, one in each of three engine rooms. The boiler rooms were placed three abreast across the width of the hull as were the engine rooms. The boiler rooms were forward of the engine rooms with a 16ft auxiliary machinery space between. The boiler uptakes to the funnel passed under the lower hangar deck in order to be protected by the armoured deck.

On trials in May 1938, she attained 31.75kts at 103,044shp and could have met her designed speed of 30kts at ¾ power. The electrical plant consisted of six 400kw steam powered generators, one in each boiler room and the other three on the main deck on the port side; two diesel powered generators in the initial design were not fitted. Aviation fuel capacity was 100,000 gallons stored in two large tanks immediately forward of the boiler rooms and sixteen smaller tanks immediately aft of the engine rooms. Fuel was delivered from the tanks via an air pressure system eliminating the fuel contamination prevalent in earlier water pressure systems.

Eight twin dual purpose 4.5in Mk III guns were fitted on sponsons at the corners of the flight deck where they had excellent fields of fire. A limited ability to fire across the deck was little used on account of the blast damage it caused. Each pair of mounts had a dedicated director, which could control all the guns on the same side of the ship, or all eight mountings if required. One director was outboard of the island, the other three along the edges of the flight deck.

Eight Mk III 0.5in multiple machine guns were mounted in pairs on separate sponsons forward of each 4.5in gun battery. Four Mk VI octuple pompoms were fitted to starboard: two forward of the island on the flight deck, one on the after end of the island superfiring over another on the flight deck. All the pompom directors were fitted on the island.

The overall design was considerably hampered by the artificial tonnage limit and welding was used extensively to save weight. Eventually, 65% of the structure would be welded including bulkheads, decks, the shell plating above the lower hangar deck, and the forward 100ft of the ship in its entirety. Welding was then a new construction technique and a welding shop and school was set up at the building slip. Some 7 million feet of welding rod weighing 260 tons was used, which saved 500 tons in rivets. Overall the use of welding caused few problems, but D quality steel was difficult to weld and some welded joints in the anti-splinter bulkhead failed during tests.

In spring 1936, with the ship completed up to the main deck, a proposal was made to install a single armoured hangar with 1in of armour on the flight deck. This would have involved so much re-working and consequent delays that it was dropped.

Post-trials the funnel was raised by 8ft to reduce smoke interference with the pompom directors on the island, and the platforms for the as yet unfitted port side pompoms were lowered below the flight deck level.

## *ILLUSTRIOUS, VICTORIOUS* AND *FORMIDABLE*

With the construction of *Ark Royal* well in hand, attention turned to the carrier to be ordered in the 1936 programme. With the design department swamped producing detailed drawings for *Ark Royal*, the simplest approach would have been an improved version built to the treaty limit of 23,000 tons. Instead, the new carrier became the armoured *Illustrious*.

## BUILDING DATA

| Ship | Built By | Laid Down | Launched | Commissioned |
|---|---|---|---|---|
| *Hermes* | Armstrong Whitworth | 15.1.1918 | 11.9.1919 | 19.2.1924 |
| *Ark Royal* | Cammell Laird | 16.9.1935 | 13.4.1937 | 16.11.1938 |
| *Illustrious* | Vickers Barrow | 27.4.1937 | 05.4.1939 | 25.5.1940 |
| *Victorious* | Vickers Newcastle (Walker) | 04.5.1937 | 14.9.1939 | 29.3.1941 |
| *Formidable* | Harland & Wolff (Belfast) | 17.6.1937 | 17.8.1939 | 24.11.1940 |
| *Indomitable* | Vickers Barrow | 10.11.1937 | 26.3.1940 | 10.10.1941 |
| *Implacable* | Fairfield (Clydebank) | 21.2.1939 | 10.12.1942 | 22.5.1944 |
| *Indefatigable* | John Brown (Clydebank) | 03.11.1939 | 08.12.1942 | 03.5.1944 |
| *Victorious* reconstructed | Portsmouth Naval Dockyard | 23.10.1950 | N/A | 14.1.1958 |

| Ship | Building Time | Standard Displacement | Full Load Displacement | Fuel |
|---|---|---|---|---|
| *Hermes* | 73 months | 11,850 tons | 13,700 tons | 2000 tons oil fuel, 7500 gal aviation fuel |
| *Ark Royal* | 38 months | 22,000 tons | 27,720 tons | 4620 tons oil fuel, 100,000 gal aviation fuel |
| *Illustrious* | 37 months | 23,000 tons | 28,210 tons (31,630 wartime) | 4850 tons oil fuel, 50,660 gal aviation fuel |
| *Victorious* | 46 months | 23,000 tons | 28,210 tons (31,630 wartime) | 4850 tons oil fuel, 50,660 gal aviation fuel |
| *Formidable* | 41 months | 23,000 tons | 28,210 tons (31,630 wartime) | 4850 tons oil fuel, 50,660 gal aviation fuel |
| *Indomitable* | 47 months | 23,080 tons | 29,730 tons | 4500 tons oil fuel, 75,110 gal aviation fuel |
| *Implacable* | 55 months | 23,460 tons | 32,110 tons | 4690 tons oil fuel, 94,650 gal aviation fuel |
| *Indefatigable* | 60 months | 23,460 tons | 32,110 tons | 4690 tons oil fuel, 94,650 gal aviation fuel |
| *Victorious* reconstructed | 87 months | 30,300 tons | 35,500 tons | 4180 tons oil fuel, 327,800 gal aviation fuel |

| Ship | SHP | Speed | Armour Weight | Aircraft Capacity | Complement |
|---|---|---|---|---|---|
| *Hermes* | 40,000 | 25 knots | ‹none› | 20 | 700 |
| *Ark Royal* | 102,000 | 31.75 knots | 2854 tons | 60 | 1580 |
| *Illustrious* | 111,000 | 30.5 knots | 4941 tons total, 1500 flight deck | 36 | 1229, 2000 in BPF |
| *Victorious* | 111,000 | 30.5 knots | 4941 tons total, 1500 flight deck | 36 | 1229, 2000 in BPF |
| *Formidable* | 111,000 | 30.5 knots | 4941 tons total, 1500 flight deck | 36 | 1229, 2000 in BPF |
| *Indomitable* | 111,000 | 30.5 knots | 4299 tons total, 1500 flight deck | 45 | 1392, 2100 in BPF |
| *Implacable* | 148,000 | 32.5 knots | 3645 tons | 48 | 1400, 2300 in BPF |
| *Indefatigable* | 148,000 | 32.5 knots | 3645 tons | 48 | 1400, 2300 in BPF |
| *Victorious* reconstructed | 110,000 | 29.25 knots | 3340 tons | 36 | 2400 |

*Formidable* with her forward windbreak up and wearing Camouflage Scheme 2 (see Alterations section for details). There is a Sea Hurricane on an outrigger in front of the forward pompoms. The HF/DF aerial is still on the foremast, with Type 281 at the masthead. *(Seaforth collection)*

Individual carrier tonnage was limited by treaty so effective armour protection schemes would inevitably lead to carriers with smaller air groups. The Admiralty did not want to allocate precious tonnage and equally scarce Sterling to such ships, preferring instead to build carriers which maximised aircraft capacity, such as *Ark Royal.* Three factors combined to influence a change in thinking.

Firstly, the London Naval treaty of 1936 removed the overall tonnage limit for aircraft carriers; the Royal Navy could now build as many carriers to the individual treaty limit of 23,000 tons as it could afford. Smaller air groups would be less of an issue with many carriers available.

Secondly, the worsening international situation brought increases in the defence budget for re-armament. Thirdly, and perhaps the most important, in 1936 the Third Sea Lord and Controller of the Navy was Rear Admiral Sir Reginald Henderson. The ranking naval aviation expert in the Royal Navy, he had been the first Rear Admiral Aircraft Carriers in 1931 and had played an important role in the development of carrier operations. He considered that *Ark Royal* was vulnerable to air attack and pressed aggressively for the armoured carrier.

Perceiving the need for the armoured carrier to be extremely urgent, in January 1936 Henderson turned the design directly

*Illustrious* working up in August 1940. The extension to the bridge wind deflector has not been fitted. The safety barrier stanchions are raised; the barrier is not rigged. The hull openings under the forward 4.5in mounts are submarine lookout positions. The round opening to port under the forward round down is the foghorn. *(Seaforth collection)*

over to the head of the Aircraft Carrier Design department, W A D Forbes, who had designed the *Ark Royal*. Henderson wanted the hangar of the new ship to be armoured against 500lb bombs dropped from 7000ft, 1000lb bombs from under 4500ft, and cruiser shellfire from outside 7000yds; the ship had to remain in action after damage rather than having to retire for repairs. Overall dimensions, speed, machinery and magazine protection, and armament, were to be similar to *Ark Royal*.

This was a very complex set of specifications and to get it all to fit within the 23,000-ton limit was no easy task. Due in no small part to the very close and informal working relationship between Henderson and Forbes, and despite being very busy with detailed drawings for *Ark Royal*, Forbes had the first design studies done by February. In late June a full sketch design was forwarded to the Naval Staff where it received approval within a month; detailed design was finished and approved in December 1936. The speed was remarkable given that the usual design process took over two years.

The design featured an armoured box over the machinery and magazines, and an armoured hangar box with the lifts at the forward and after ends placed outside of the hangar. Placing heavy armour high in the ship required reducing the freeboard in order to maintain stability; only a single hangar capable of storing 36 aircraft could be fitted. The design was subject to extensive aerodynamic testing to minimise airflow disturbance over the flight deck, the flight deck featuring round downs fore and aft and the island was aerofoil shaped.

Armour plate is usually affixed to existing hull plating, but there was not enough weight available for this to be done with the armoured flight deck. It was instead fitted as a structural member with no backing plate. This was a very noteworthy technical achievement as it had never been done before with a flight deck. Under the terms of the Washington Treaty, Britain was obliged to report the standard displacement and numbers of aircraft of the new carriers four months prior to the ships being laid down. This would reveal that the

*Illustrious* on 24 November 1941 at the end of her Norfolk repair freshly repainted into Camouflage Scheme 1 (see Alterations section for details). The flight deck aft has been raised and the wind deflector fitted around the front of the bridge. 'A' director has been fitted with Type 285 radar. Single 20mm mounts can be seen in the new sponsons on the island. *(US National Archives 80-G-15741 via Roger Torgeson)*

*Illustrious* in 1947 with multiple 40mm Boffins and 40mm single guns along the edges of the flight deck and in the island galleries. Radar type SM1 has displaced the HADT atop the bridge which has been moved to the flight deck in place of S2 pompom. What appears to be a crane or gantry is stowed facing aft just forward of the 4.5in mounts. *(Seaforth collection)*

ships were 1000 tons heavier than *Ark Royal* while carrying half the number of aircraft, revealing that the ships were probably heavily armoured. To keep the secret for as long as possible, the aircraft capacity was reported as being dependent on the types embarked, with capacity for 60. This was true if a deck park was used, but at that time the Royal Navy never anticipated using one. An interesting bending of the rules!

The British armour plate industry had all but vanished under the weight of disarmament, depression, and the lack of new orders; output was down to 18,000 tons annually in 1936. Orders for armour plate for the flight decks of *Illustrious*, *Victorious*, and *Formidable* were therefore placed with Virkovice of Morava Ostrava in Czechoslovakia. The last plates were delivered via train across Germany in August 1939.

In June 1936, the Defence Policy & Requirements committee approved the Admiralty proposal for six new carriers, two in the 1936 programme. Tenders went out in November 1936 with Vickers-Armstrong being the successful low bidder. Orders were placed for both ships (*Illustrious* and *Victorious*) on 13 April 1937.

Two repeats were ordered in the 1937 programme (*Formidable* and *Indomitable*). The last pair were to be ordered in 1938, but with government finances rationed between all three services and with fears that war with Germany was more likely than war with Japan, this was reduced to one in the 1938 programme (*Implacable*), and one in 1939 (*Indefatigable*). The first of the class, HMS *Illustrious*, was commissioned on 25 May 1940.

With a distinctive aerofoil shape, the 170ft long island was made up of four decks. The wheelhouse was on the lowest level deck 'A', with the compass platform and open navigation position on deck 'B' above. The Admiral's bridge and viewing platform was on deck 'C' and the signal bridge was on deck 'D' at the top. To counterbalance its weight and that of the armoured boiler uptakes, the oil fuel tanks on the port side within the torpedo protection system were 15ft wider than those to starboard.

An improved homing beacon designed to work with other electronic gear, the Type 72DM, was fitted on the tripod mast. Above it was a High Frequency Direction Finder loop and a radar receiving aerial. The radar transmitting aerial was fitted to a mast on the flight deck at the after end of the island. *Illustrious* had her completion date pushed back by two months waiting for this radar equipment to be installed.

The flight deck was constructed of 3in armour plate between the lift wells, the armour plate extending forward and aft halfway along the length of each lift well. The rest of the deck immediately around the lift wells was constructed of 1.5in high tensile steel. The flight deck forward and aft of the lift wells was not armoured.

A single BH3 catapult was fitted on the port side forward. To avoid any cuts into the armoured deck, it was mounted atop the deck within a raised casing with bevelled edges to allow aircraft to taxi over it. Nine

*Victorious* coming alongside at Pearl Harbor on 3 March 1943. Directly below the 20mm gallery, the three rows of her Signal Group lights for night flying can be seen. The YE aerial has been fitted and the aftermost director has Type 285. The Type 272 radar lantern can be seen above the bridge. The ship is wearing Camouflage Scheme 2 (see Alterations section for details) and still has 8 antenna masts. The openings under the forward 4.5in mounts are submarine lookout positions. *(Author's collection)*

*Victorious* in Hvalfjord, Iceland on 4 October 1941. *(US National Archives 80-G-463650 via Roger Torgeson)*

arrester wires spaced 24ft apart were fitted aft, each capable of stopping 20,000lbs within 150ft; the gear could be set for particular aircraft types. *Victorious* and *Formidable* were fitted with two additional wires forward for landings over the bow. Two safety barriers that could stop 20,000lbs at 40kts within 62ft were installed abreast the island forward of the arrester wires. Eight antenna masts were fitted along the edges of the flight deck.

The fully enclosed single hangar was situated between the lift wells and measured 456ft x 62ft with a height of 16ft and could be split into three sections by two metal fire-proof curtains which rolled down from the deckhead. The number of aircraft that could be stowed depended on size and type, but it was designed to hold 36.

Along its sides were 4.5in armoured bulkheads, closed off at the lift wells by 4.5in armoured transverse bulkheads. The roof of the armoured hangar box was formed by the 3in flight deck. There was no deck armour anywhere within the hangar; 1in anti-splinter protection covered the portions of the deck that extended forward and aft of the side belt.

The single platform unarmoured lifts were fitted at each end of the hangar where they did not take up any aircraft stowage space. Access to the hangar was through 4.5in armoured shutter doors fitted into the transverse armoured bulkheads. Electrically operated, both lifts measured 45ft by 22ft and could lift 13,440lbs with a cycle time from hangar to flight deck of 30 seconds. A roll-down fire curtain was fitted in each lift well outside the armoured shutters. There was a torpedo and bomb lift on

A weather-beaten *Formidable* in Sydney Harbour on 24 August 1945. Note the 40mm Boffin at the front edge of the flight deck, Type 277 on the top of the bridge, YE mast and aerial attached to the funnel, the HADT director on the flight deck in front of the bridge, and the letter 'R' at the bows. She is wearing Camouflage Scheme 4 (see Alterations section for details). *(By courtesy of David Hobbs)*

## DIMENSIONS

| Ship | Length (oa) | Length (wl) | Beam (max) | Beam (wl) |
|---|---|---|---|---|
| *Hermes* | 600ft | 578ft 5in | 95ft 9in | 70ft 3in |
| *Ark Royal* | 800ft | 725ft | 112ft | 94ft 10in |
| *Illustrious* | 740ft – 1940<br>748ft 6in – 1945 | 710ft | 106ft 9in | 95ft 9in |
| *Victorious* | 745ft – 1940 | 715ft | 106ft 9in | 95ft 9in |
| *Formidable* | 740ft – 1940 | 710ft | 106ft 9in | 95ft 9in |
| *Indomitable* | 753ft 11in | 710ft | 116ft 3in | 95ft 9in |
| *Implacable* | 766ft 2in | 730ft | 141ft 6in | 95ft 9in |
| *Indefatigable* | 766ft 2in | 730ft | 141ft 6in | 95ft 9in |
| *Victorious* reconstructed | 778ft 3in | 710ft | 145ft 9in | 103ft 4in |

| Ship | Flight Deck | Hangar(s) | Lifts | Freeboard to flight deck |
|---|---|---|---|---|
| *Hermes* | 570ft x 90ft | 400ft x 50ft x 16ft | 46ft x 47ft<br>46ft x 47ft | 32ft |
| *Ark Royal* | 720ft x 95ft Excluding round down<br>797ft x 95ft with round down | 568ft x 60ft x 16ft (upper)<br>452ft x 60ft x 16ft (lower) | 45ft x 25ft<br>45ft x 22ft<br>45ft x 22ft | 59ft 11in |
| *Illustrious* | 620ft x 95ft – 1940<br>670ft x 95ft – 1942<br>695ft x 95ft – 1943<br>740ft x 95ft – 1944 | 456ft x 62ft x 16ft | 45ft x 22ft<br>45ft x 22ft | 38ft |
| *Victorious* | 620ft x 95ft – 1941<br>695ft x 95ft – 1943<br>740ft x 95ft – 1944 | 456ft x 62ft x 16ft | 45ft x 22ft<br>45ft x 22ft | 38ft |
| *Formidable* | 620ft x 95ft – 1940<br>629ft x 95ft – 1943<br>740ft x 95ft – 1944 | 456ft x 62ft x 16ft | 45ft x 22ft<br>45ft x 22ft | 38ft |
| *Indomitable* | 745ft x 95ft | 416ft x 62ft x 14ft (upper)<br>168ft x 62ft x 16ft (lower) | 45ft x 33ft<br>45ft x 22ft | 52ft |
| *Implacable* | 760ft x 90ft | 456ft x 62ft x 14ft (upper)<br>208ft x 62ft x 14ft (lower) | 45ft x 33ft<br>45ft x 22ft | 50ft |
| *Indefatigable* | 760ft x 90ft | 456ft x 62ft x 14ft (upper)<br>208ft x 62ft x 14ft (lower) | 45ft x 33ft<br>45ft x 22ft | 50ft |
| *Victorious* reconstructed | 775ft x 145ft<br>8deg angle | 412ft x 62ft 6in x 17ft 6in (includes 52 ft extension fwd of forward lift) | 58ft x 40ft<br>54ft x 34ft | 42ft 1in |

the port side amidships and a bomb lift forward on the starboard side under the island; both opened into the hangar and onto the flight deck.

The armoured box comprised a 4.5in armour belt amidships 295ft long, which extended from the lower deck to the upper deck, stepping up to the hangar deck abreast the magazines and up to the gallery deck amidships on the starboard side to cover the funnel uptakes. At the hangar deck it met 3in of deck armour that extended inboard to the hangar bulkheads. The armoured box was closed off by 2.5in armoured transverse bulkheads extending across the ship at the point where the belt armour stepped up to the hangar deck, protecting the machinery, magazines, and aviation fuel tanks.

The steering gear was within a separate armoured box with 3in top and sides and 2.5in bulkheads. The torpedo protection ran the length of the side belt and was the same scheme of air-liquid-air compartment 'sandwich' backed by a 1.5in anti-splinter bulkhead that had been used for *Ark Royal*.

The island sides and signal deck were protected by bullet proof plating. The weight of armour in *Illustrious* was 4491 tons, double the machinery weight and 21% of total displacement.

The same weight and space issues described for *Ark Royal*'s power plant also applied to the armoured carriers. In consequence the same arrangement of turbines, boilers, engine rooms, and boiler rooms powering three propeller shafts was fitted in the new ships; the space between the boiler and engine rooms was increased to 36ft. The three-shaft arrangement led to serious vibration problems and propellers were changed and modified numerous times. *Illustrious* attained 30.5kts at full power of 111,000shp. With one boiler room shut down reducing power to 92,500shp, she hit 29kts and could make 25kts at half power, 55,000shp. During trials some vibration was noted in the island, but was not deemed serious. Later in the war, *Illustrious* would suffer badly from vibration problems.

A total of 2400kw of electric power was generated by six 400kw steam generators;

one in each wing engine room and four more in separate compartments on the main deck within the armoured box. As with *Ark Royal*, no auxiliary diesel generators were fitted.

There was stowage for 50,450 gallons of aviation fuel in 20 tanks on the hold deck in six compartments. Two of the compartments were between the boiler and engine rooms, the others were forward of the boiler rooms. The amount of fuel was half of that carried aboard *Ark Royal* and it was estimated that *Victorious* had only 78% of the fuel required for one month of flying time for her air group – in 1938! It would prove to be hopelessly inadequate later in the war during sustained operations.

Eight twin dual purpose 4.5in Mk III BD gun mounts were placed in pairs on sponsons at the corners of the flight deck, the sponsons having to be quite deep as a great deal of each mount was actually below deck. The tops of the mounts protruded 2ft 6in above the flight deck to allow for cross deck firing. One HADT director was atop the island, the other three along the edges of the flight deck.

Six Mk VIA* octuple pompoms were fitted: four to starboard with the same arrangement as *Ark Royal*, and two on port side sponsons amidships. Screens were fitted to S1 and S2 to preserve the smooth flow of air around the island. Five of the directors were on the island, the sixth on a sponson on the port side just forward of P1 pompom.

*Illustrious* was equipped with a 6¾-ton mobile crane nicknamed 'Jumbo' that could lift 2.5 tons.

## INDOMITABLE

The return of full control of its air assets to the Admiralty via the Inskip Award of 1937 prompted a fresh re-appraisal of the small air groups carried by the armoured carriers. Increasing the size of the air group would require a return to the double hangars of *Ark Royal*. The design changes would be considerable.

Finding the necessary weight to fit a second hangar above the first one required reducing the hangar side bulkhead armour thickness to 1.5in, and the outboard hangar deck to 2.5in. The flight deck remained at 3in, the armour covering the same extent of the flight deck as in the initial design.

While both hangars had 1.5in armour on the side bulkheads, the upper hangar deck was built above the armoured box and had no other side or deck protection. The side belt protection remained the same as the earlier carriers, with the side belt coming up to the lower hangar deck amidships where it met the 2.5in outboard lower hangar deck.

The existing hangar, now the lower hangar, had to be reduced in length to provide aircraft workshops and accommodation for the increased numbers of aircrew. The new upper hangar was shorter than that of the earlier carriers at 416ft x 62ft and 14ft high; the lower was 168ft x 62ft and 16ft high; both hangars shared a common after bulkhead. The new hangar arrangements boosted her capacity to 45 aircraft.

A larger forward lift of 45ft x 33ft was fitted while the after lift remained at 45ft x 22ft; both were single platforms. The widened forward lift allowed fixed wing aircraft such as the Seafire and Sea Hurricane to be struck below. The forward lift served only the upper hangar, the after lift served both hangar decks. To accommodate the shortened length of the upper hangar, the forward lift well was moved aft

A very weather-beaten *Indomitable* following HMS *Warspite* in May 1942. There are two 20mm single mounts at the aft corner of the flight deck. Note the Sea Hurricanes on deck, the ship is wearing Camouflage Scheme 1 (see Alterations section for details). *(Author's collection)*

*Indomitable* at the completion of her Norfolk refit in March 1944, wearing Camouflage Scheme 3 (see Alterations section for details). Note the stern AA gallery with six 20mm, quarterdeck AA gallery with three 20mm, and the new AA galleries outboard of the island. The mobile crane is parked right forward and two yard cranes are on the flight deck aft. The mast outboard of the island is for the YE beacon, and an SM1 radar can be seen at the front of the island. *(Author's collection)*

*Implacable* in mid-1944 in her first camouflage scheme. Visible on the flight deck are Fireflies and Avengers. *(By courtesy Vancouver Maritime Museum)*

16ft and the after lift well moved forward 24ft.

Nine arrester wires were fitted aft and three forward, the after wires being spaced 20ft apart to accommodate the reduced distance between the lifts. The forward wires were for emergency landings over the bow. Three safety barriers were fitted. The single catapult was upgraded to 12,500lbs at 66 knots. Aviation fuel stowage was increased by 25,000 gallons.

The first three armoured carriers were too far along in construction for such major changes but *Indomitable* was facing an eight-month delay pending delivery of her armour plate. Adding the extra hangar would not appreciably delay her overall construction time. She completed seven months after *Victorious*. Reflecting ongoing improvements, the ship did not have the after round down, her initial radar suite was considerably enhanced, and she was fitted with eight single 20mm guns. The increased freeboard made her a better sea boat than the earlier armoured carriers and her 16ft high lower hangar could accommodate the American built aircraft that would soon enter RN service. She was considered to be the best of the armoured carriers.

## *IMPLACABLE, INDEFATIGABLE*

Construction was suspended on both ships in 1940 and not resumed until 1942. The delay allowed a further redesign to include double hangars, a wider and longer flight deck, and a speed increase to 32kts.

The three-shaft power train could not deliver the required horsepower to attain 32 knots. The ships therefore reverted to a more traditional four-shaft arrangement, with an extra boiler room and engine room worked into the hull, which was lengthened by 26ft to provide the necessary room. The four boiler rooms were directly under the island, arranged two to port and two to starboard. The wing engine rooms driving the outboard shafts were directly abaft the boiler rooms and the centre engine rooms for the inboard shafts were immediately aft of those. Diesel generators for emergency electrical power were also fitted.

They used the same abbreviated armour scheme and layout as *Indomitable* in addition to which the transverse hangar bulkheads were reduced to 2in and the side belt increased in length to 304ft to cover the larger machinery spaces; the upper hangar remained outside the armoured box. The

magazines were re-arranged, with the aircraft ordnance magazines within the armoured box and the 4.5in magazines outside but directly protected with 2in to 3in crowns and 1.5in to 2in bulkheads. A proposal to increase the hangar side protection to 2in resulted in the height of both hangars being set to 14ft. The ships were completed with 1.5in hangar sides in any case.

The upper hangar was the same dimensions as *Illustrious*, 456ft x 62ft. The lower was 208ft x 62ft as that deck also provided accommodation for the increased numbers of machinery personnel in addition to aircrew and workshop space; both hangars shared a common after bulkhead.

The flight deck was widened to port forward of the after 4.5in guns. The after round down was eliminated and the flight deck featured a less prominent rounded edge at the bow. The 4.5in turrets were fitted with flat roofs which were level with the flight deck.

The two lifts were the same dimensions as those on *Indomitable* with their flight deck positions reverting to those on *Illustrious*. The forward lift served only the

*Implacable* passing under the Lions Gate bridge outbound from Vancouver on 18 October 1945. All of the port side installations can be made out and there is a single 20mm visible in the hull opening just ahead of the forward 4.5in guns. The ship is still wearing her Admiralty Standard camouflage scheme; note the coloured band along the base of the island. The ship's crest with battle honours is on the side of the island. Despite the wet conditions, the crew is manning the side and the band is performing. *(Author's collection)*

*Indefatigable* in March 1946 wearing the Dark Hull Light Upper Works paint scheme. Note the 40mm Boffin below S1 pompom and the Royal Marine contingent drawn up on the flight deck. After this date she will no longer be an operational carrier. *(By courtesy of Nigel Denchfield)*

upper hangar, and after lift served both hangars; both were rated at 20,000lbs. Eight arrester wires were fitted aft; three safety barriers were fitted amidships. The single catapult was upgraded to launch 20,000lbs at 56 knots. Outriggers were fitted to enable a larger deck park. Aviation fuel capacity was increased to 94,650 gallons.

The island was enlarged and extended to accommodate the larger boiler uptakes and wider funnel. The ships completed with an enhanced radar suite and two of the HADT were in new positions: 'A' was on the flight deck forward of the island and 'Y' was fitted on the island aft of the funnel.

Only five octuple pompoms were fitted – two forward of the island on the flight deck, one aft of the island on the flight deck, and two on port side sponsons. A quad pompom was fitted on the port side just aft of the octuple pompoms. The light AA was also considerably enhanced.

## THE MODERNISATION OF *VICTORIOUS*

Plans were drawn up in 1947 to modernise all of the *Illustrious* class. The ships would be stripped down to the hangar deck and rebuilt incorporating a 17ft6in high hangar with the armoured flight deck being retained, a new island fitted, and cabling and wiring renewed. The 4.5in guns would be replaced with the new 3in/70 under development. Each ship would take three years to complete at an estimated cost of £2.5 million each, much less than the estimated £7.0 million to build a new ship. Each modernised ship would be capable of operating the newer and heavier jets and would have 20 years of hull life left.

*Formidable* and *Victorious* were approved for reconstruction in 1948 with *Implacable*, *Indefatigable*, and *Indomitable* to follow; *Illustrious* would serve as deck landing training ship. The detailed design was ready in June 1950 at which time *Formidable* was surveyed and found to be in such poor condition that *Victorious* was taken in hand instead, the work to be done at Portsmouth. By then the estimated cost had risen to £5.4 million.

What was meant to be a fairly straightforward reconstruction suffered from being done at a time of considerable innovation. The plans and drawings would have to be continually recast to incorporate three major post-war developments: the steam catapult, the angled flight deck, and the Type 984 radar system with its associated operations spaces. Further complications arose over storage space for guided missiles, nuclear weapons, and workshops for the increasingly complex airborne radar systems.

At the time it was the largest project ever conducted at a Royal Dockyard and the work exposed many inadequacies in project planning. A shortage of shipwrights on site meant that fabrication of the island and gallery deck had to be parcelled out to Devonport Dockyard.

The angled flight deck was approved in 1953. In December 1953 it was determined that the original steam plant was worn out

*Victorious* in her first commission post modernisation with Skyraiders, Scimitars, and Sea Venoms on deck. The retractable catapult control position can be seen forward of the lift well; the jet blast deflectors are outlined just aft of the lift well. The mobile crane 'Jumbo' is at the far right. *(Author's collection)*

*Victorious* at speed with water pouring out of her forward hull openings. On deck are Buccaneers and Sea Vixens. The sponson for the Deck Landing Projector can be seen with a Wessex helicopter just beyond. *(Author's collection)*

and not sufficient to power both the ship and the steam catapults. The already completed flight deck had to be partially dismantled in order to replace the boilers. The Type 984 radar required wiring upgrades and increased electrical capacity. Costs rapidly ballooned from the original estimate to over £30 million by the time the ship was completed in January 1958. The soaring costs and extended project time line resulted in the cancellation of modernisation plans for the rest of the class; those ships were sent for scrap instead. Despite the issues, when *Victorious* re-commissioned in January 1958, she was a fully modern ship capable of operating the current generation of naval jets within a vastly different threat environment.

The hull was fitted with 4ft bulges on either side for stability and to allow a wider hangar and upper decks; a 2in armour belt was fitted outside of the bulges. The extra width prevented her from using the Panama Canal. The hangar side armour was reduced to 1.5in, and the deck outside of the hangar was reduced to 2.5in. The axial portion of the flight deck remained at 3in; most of the angled portion was not armoured. The additions brought the deep load displacement to 35,500 tons and the ship actually floated 4ft lower in the water. This precluded the idea of a deck edge lift as it would have been too close to the water.

The flight deck was angled at 8¾ degrees to port over a large sponson and was equipped with four arrester wires. A gyro stabilised Deck Landing Mirror Sight was fitted on both sides of the flight deck. Two steam catapults were fitted forward with blast deflectors aft of each one; the catapults were capable of accelerating 40,000lbs to 105kts. A retractable control position was fitted between them.

Two enlarged and non-armoured lifts were fitted. The one aft was 54ft x 34ft and was fitted in the same position as the original. The forward measured 58ft x 40ft and was 36ft aft of the original. Both could lift 13 tons at normal speeds and 17 tons at slow speed. The main hangar was 360ft x 62ft with a 56ft extension forward of the forward lift bringing the overall length to 416ft.

The electrical capacity was upped to 5000kw with eight steam and four diesel generators. The island size was reduced to 130ft in length and only three decks with a Flying Control position ('Flyco') fitted on the port side forward. As part of nuclear, biological, and chemical defences, the compass platform on deck 'B' was made airtight.

The 30-ton Type 984 radar unit was fitted atop the island. It was the best air defence radar system in the world at the time, providing air warning out to 210 miles and fighter direction to 90 miles. A Type 957 Tacan navigation beacon and Type 293Q target indication radar were fitted to the lattice mast. Type 963 approach radar was fitted in a radome at the after end of the island and a Type 978 navigational radar was fitted atop the bridge. A hinged mast forward on the starboard side carried a VHF/DF antenna, and there were four antenna masts on each side of the after flight deck.

The 3in/70 mount was not ready so the ship was fitted with six American Mk 33 twin 3in/50 mounts with enclosed shields and a Mk 34 radar; three each port and starboard. One six-barrel Mk VI 40mm was fitted on the starboard side abaft the boat crane.

## SELECTED OPERATIONS 1939–1945

Despite the underlying issues with aircraft, ships, and doctrine, it must not be forgotten that the RN excelled at carrier operations in spite of it all. They pioneered the concepts of radar controlled fighter direction, multiple carrier task forces, strikes against enemy fleets in port, strikes against enemy fleet units at sea to damage and slow them down for the battle fleet, strikes against strategic shore installations, and operations in support of land forces. Their pilots and deck crews were consummate professionals, achieving spectacular results while often operating slow and obsolescent aircraft.

The RN carriers were in action from the first day of the war until the last; the following is a list of the more significant operations undertaken by the carriers described in this book.

**02.10.1939 to 15.2.1940,** *Ark Royal,* Operations in the South Atlantic in search of the *Graf Spee*.

**24.4.1940 to 13.6.1940,** *Ark Royal,* Operations off Norway, often in company with *Glorious* operating as a carrier task force.

**13.6.1940,** *Ark Royal,* Strike against *Scharnhorst* in Trondheim, Norway; 8 out of 15 Skuas were shot down in exchange for one bomb hit which did not detonate.

**03.7.1940 to 06.7.1940,** *Ark Royal,* Operations against the French fleet at Mers-el-Kebir (Operation Catapult).

**08.7.1940,** *Hermes,* Attack on French battleship *Richelieu* at Dakar (Operation Catapult).

**01.8.1940 to 13.11.1941,** *Ark Royal,* 'Club' runs to Malta delivering aircraft, in company at various times with *Argus*, *Furious*, and *Victorious*. Delivered 250 Hurricanes, 14 Fulmars, 8 Swordfish, and 11 Albacores over the course of 14 operations.

**23.9.1940 to 25.9.1940,** *Ark Royal,* Covered attempted landings at Dakar (Operation Menace).

**11.11.1940,** *Illustrious,* Attack on Taranto – the Fleet Air Arm's most celebrated success. Italian battleships *Littorio* and *Duilio* were heavily damaged, *Conte de Cavour* sunk and never returned to service (Operation Judgement).

**27.11.1940,** *Ark Royal,* Battle of Spartivento with strikes against the retreating Italian fleet, no hits were scored.

**09.2.1941,** *Ark Royal,* Strikes against shore targets and spotting for the battle fleet during the bombardment of Genoa.

**28.3.1941,** *Formidable,* Battle of Matapan supporting the battle fleet. Torpedo hits scored on battleship *Vittorio Veneto* and heavy cruiser *Pola* bringing the latter to a standstill and leading directly to her loss along with two sister ships at the hands of the battle fleet.

**.4.1941 to .6.1941,** *Hermes,* Operations in the Persian Gulf against pro-Axis forces in Iraq.

**24.5.1941,** *Victorious,* Strike against *Bismarck*, 1 torpedo hit amidships.

**26.5.1941,** *Ark Royal,* Strike against *Bismarck*, 1 torpedo hit amidships, 1 torpedo hit aft disabling the ship's steering gear.

**30.7.1941,** *Victorious,* Strike against Kirkenes and Petsamo, Norway in company with *Furious*; 12 Albacores and 2 Fulmars lost.

**27.1.1942 to 28.1.1942,** *Indomitable,* Flew off 50 RAF Hurricanes to Batavia in Java.

**09.3.1942,** *Victorious,* Covering Russian convoy PQ12, strike launched against *Tirpitz* while she was at sea, no hits were scored.

**.3. to .4.1942,** *Hermes*, *Formidable*, *Indomitable,* Operations in the Indian Ocean against Japanese carrier fleet; contact was never made between the two forces.

**05.5.1942 to 08.5.1942,** *Illustrious*, *Formidable*, *Indomitable,* Covered landings in Madagascar at Diego Saurez (Operation Ironclad).

**09.8.1942 to 12.8.1942,** *Victorious*, *Indomitable,* Covered the passage of a convoy to Malta in company with *Argus*, *Eagle*, and *Furious* (Operation Pedestal).

**08.11.1942 to 10.11.1942,** *Victorious*, *Formidable,* Covered the landings in North Africa. Four Martlets from 882 Sqn from *Victorious* took the surrender of Blida airfield (Operation Torch).

**27.6.1943 to 25.7.1943,** *Victorious,* Covered landings on New Georgia in company with USS *Saratoga* (Operation Toenails).

**.7.1943,** *Formidable*, *Indomitable,* Covered landings on Sicily (Operation Husky).

**.9.1943,** *Illustrious*, *Formidable,* Covered landings in Italy at Salerno (Operation Avalanche).

**03.4.1944,** *Victorious,* Strikes against *Tirpitz* in Kaafjord, Norway in company with *Furious* (Operation Tungsten).

**19.4.1944,** *Illustrious,* Strike against Sabang, Sumatra in company with USS *Saratoga* (Operation Cockpit).

**17.5.1944,** *Illustrious,* Strike against Surabaya, Java in company with USS *Saratoga* (Operation Transom).

**19.6.1944,** *Illustrious,* Strike against Port Blair, Andaman Islands (Operation Pedal).

**17.7.1944,** *Formidable*, *Indefatigable,* Strikes against *Tirpitz* in Kaafjord in company with *Furious* (Operation Mascot).

**25.7.1944,** *Illustrious*, *Victorious,* Strike against Sabang in support of shore bombardment by fleet units (Operation Crimson).

**03.8.1944 and 10.8.1944,** *Indefatigable,* Anti-shipping strikes in Norwegian waters (Operations Turbine and Offspring).

**24.8.1944,** *Victorious*, *Indomitable,* Strikes against Padang and Emmahaven, Java (Operation Banquet).

**22.8.1944 to 29.8.1944,** *Formidable*, *Indefatigable,* Strikes against *Tirpitz* in Kaafjord in company with *Furious* (Operation Goodwood).

**18.9.1944,** *Victorious*, *Indomitable,* Strike against railway yards at Sigli, Sumatra (Operation Light).

**17.10.1944 to 19.10.1944,** *Victorious*, *Indomitable,* Strikes on the Nicobar Islands (Operation Millet).

**18.10.1944 to 09.12.1944,** *Implacable,* Anti-shipping strikes in Norwegian waters.

**20.12.1944,** *Illustrious*, *Indomitable,* Strike against Belawan Deli, Sumatra (Operation Robson).

**04.1.1945,** *Victorious*, *Indomitable*, *Indefatigable,* Strikes against oil refineries at Pangkalan Brandan, Sumatra (Operation Lentil).

**24.1.1945,** *Illustrious*, *Victorious*, *Indomitable*, *Indefatigable,* Strikes against Pladjoe oil refinery, Palembang, Sumatra (Operation Meridian One).

**29.1.1945,** *Illustrious*, *Victorious*, *Indomitable*, *Indefatigable,* Strikes against Songei-Gerong oil refinery at Palembang, Sumatra. Most devastating attack by the FAA during the war (Operation Meridian Two).

**26.3.1945 to 25.5.1945,** *Illustrious*, *Victorious*, *Indomitable*, *Indefatigable*, *Formidable,* Sustained operations against the Sakishima Gunto islands as part of the invasion of Okinawa (Operation Iceberg). *Formidable* replaced *Illustrious* on 14 April.

**12.4.1945,** *Illustrious*, *Victorious*, *Indomitable*, *Indefatigable,* Strikes against Formosa.

**14.6.1945 to 15.6.1945,** *Implacable,* Strikes against Truk, Caroline Islands (Operation Inmate).

**17.7.1945 to 15.8.1945,** *Formidable*, *Victorious*, *Implacable*, *Indefatigable,* Strikes against Japan. Lt Robert H Gray RCNVR was awarded a posthumous Victoria Cross for sinking Japanese destroyer escort *Amakusa* on 9 Aug. *Indefatigable* remained in the operating area after departure of the rest of the carriers on 12 Aug and carried out two further days of strikes.

**15.6.1953,** *Illustrious*, *Indomitable*, *Implacable*, *Indefatigable,* Coronation Fleet Review at Spithead.

# Model Products

Royal Navy carriers have never been as popular modelling subjects as their USN and IJN counterparts, which accounts for the relative scarcity of injection moulded kits up until 2015 when the Eagle, Revell and Airfix offerings from the 1960s were joined by new releases from Aoshima, Flyhawk, and Trumpeter. The new kits reflect the state of the art as regards detail, accuracy, and ease of assembly.

The 3D printing process continues to revolutionise the hobby, making possible the creation of exquisite miniature reproductions of boats, fittings, and weapons with detail and precision that exceeds their injection moulded equivalents. The selection of photo-etch accessories is also extensive with equivalent precision and detail.

## BANDAI *ILLUSTRIOUS* 1:2000 Scale

This clip together kit depicts *Illustrious* in her mid-war appearance with the bridge wind deflector and Type 72DM homing beacon. The instructions are printed on the underside of the box lid. The model can be completed as either waterline or full hull; a display stand with the ship's name is included for the full hull version. The 4.5in guns, antenna masts, directors, cranes, and foremast are separate parts. Two single-piece Swordfish make up the air group. The kit is well detailed considering its small size, with the single-piece flight deck having moulded-in arrester gear, barriers, catapult track, lifts, pompoms, and flight deck markings.

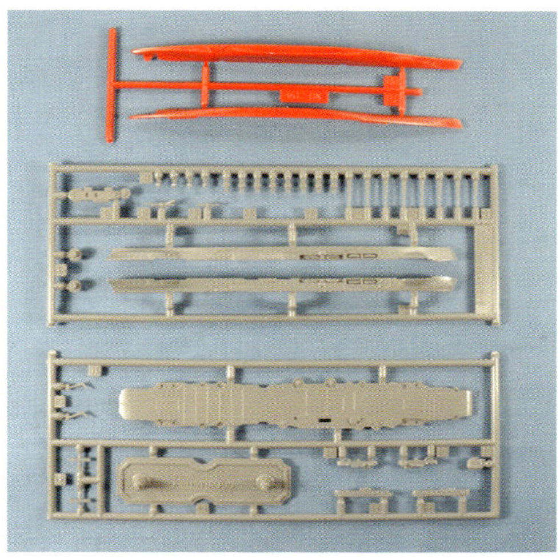

■ Review model courtesy of **Neil Ramage**

## AIRFIX *ARK ROYAL* 1:1200 Scale

This is a clip together kit in the Airfix Naval History Series. The single page double-sided foldout instructions have a simple painting guide for the ship with references to the Humbrol range.

The waterline hull and flight deck are single pieces with the antenna masts, 4.5in mounts, directors, inboard bulkheads, and

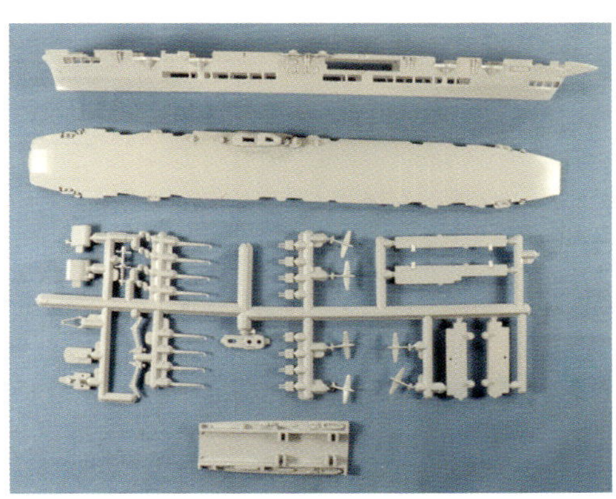

cranes all separate. The island is made up of five parts but lacks the forward rangefinder platform. The air wing consists of six single-piece Fulmars with spread wings. There is no detail on any and the undercarriages are simple nibs; there are no propellers. The hull shows considerable moulded-in detail, but there is not much detail on the kit parts, although the antenna masts and cranes have moulded-on lattice. There is no armour belt, degaussing cable, funnel cap, forecastle or quarterdeck pieces, forecastle opening under the bow, catapult loading-on platforms, or forward windbreaks, but for such a small scale, this is a tidy and crisply moulded kit, although the clip together construction will leave gaps where the flight deck joins the hull.

## EAGLE/PYRO ***VICTORIOUS*** 1:1200 Scale

The kit depicts a mid-war *Victorious* with the after round down mostly eliminated. The instructions are on a single double-sided sheet with historical notes on one side. Eagle also released the kit as *Formidable* and it was also released as part of the Table Top Navy series by the US company Pyro. Once highly affordable, it now commands high collector prices when it can be found.

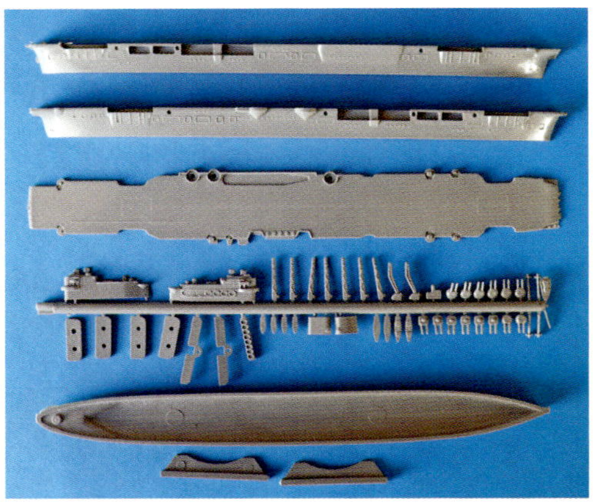

No aircraft are included, but the kit sports a lot of detail. Considering its age, this is still a good kit with just a few issues. The pompoms do not bear any resemblance to the real thing, and there are just seven antenna masts. The most glaring problem is that the island and its supporting sponson don't line up; fully half the island is aft of the sponson, the correction of which will require some complex modification.

Review model courtesy of **Jeffrey Stevenson**

## REVELL ***ARK ROYAL*** 1:720 Scale

First released in 1967 as part of the International Series, this venerable kit is still in production six decades later. There are two pages of instructions with very simple painting guides for the ship and aircraft. The air group consists of eight single-piece Fulmars with some panel lines, elevator, and canopy details; the undercarriages are simple nibs and there are no propellers.

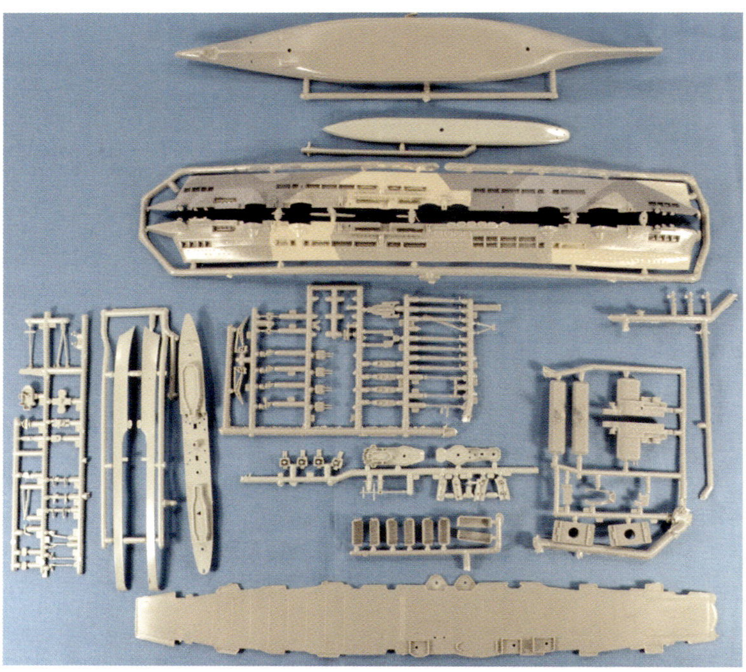

The most serious issue is at the bow where the attempt to create the forecastle opening has left a pronounced lip. There are no anchors or armour belt and the degaussing cable does not extend to the bow on either side. The four boats provided as separate parts are simplified. The lifts have squared off corners, there is neither the forward windscreen nor catapult loading-on platforms, and S2 pompom should be higher than S1. The mainmast is stepped too far forward, the canopy over the bridge is moulded as a solid structure, the platform and shield for S3 pompom is inaccurately shaped, and there are no windows on the admiral's bridge.

It is a very simple build, but the addition of some aftermarket items and tidying up the various inaccurate shapes will result in a good model. A 'Tribal' class destroyer is also included.

Review model courtesy of **Wayne Joslin**

## AOSHIMA ***ARK ROYAL*** 1941 1:700 Scale

This kit is meant to depict the ship in her 1941 appearance. The instructions are on a folded double-sided sheet with a painting and decal placement diagram for the ship and for the aircraft. Paint references are given for the Mr Color and

Aqueous Hobby Color ranges – translations in various languages will help with determining the colours. The decal sheet contains flags and aircraft and flight deck markings.

The air group is on two parts trees each containing a single Fulmar, Skua, Walrus, and folded Swordfish. The aircraft are adequately detailed with panel lines, elevators, flaps, and canopy details. One tree of the older Aoshima Royal Navy Plane set is also included.

Moulded-on detail is well executed with the hull featuring a clearly defined armour belt, well defined forward and aft hull knuckles, recessed portholes, raised edges around the hull openings, supports for overhanging decks and sponsons, boat booms, and the elevating rams for the side windbreaks forward. The island is made up of 15 parts and sports the early bridge with the rangefinder platform below the level of the open bridge. The flight deck has recessed catapult tracks and forward sheaves, the catapult loading on platforms, forward windbreak, protecting ramps forward and aft of the lifts, safety barrier, and arrester wires. The antenna masts and cranes have moulded-on lattice and the funnel cap has an open funnel screen.

There are a few minor issues. The swing-out catwalk on the port side of the island is represented by a curved bulwark. The inboard bulkheads do not have much detail. The boats are generic types and only seven are fitted – none in the aftermost galleries. The hull opening at the bow under the round down is not well-defined. The platforms for S1 and S2 pompoms are inaccurately shaped and should be omitted. To be strictly accurate for the 1941 time period, a degaussing cable needs to be fitted to the hull, the protecting ramps around the lifts removed, and splinter shields added to S1, S2, and S4 positions. Up to May 1941, the portside pompoms were not fitted.

***Ark Royal* 1939:** This is the exact same kit and components with an additional tree of boats and a revised set of flight deck markings on the decal sheet. The ship did not carry Fulmars in 1939, which will reduce the already small air group. The port side pompoms and the shield inboard of S3

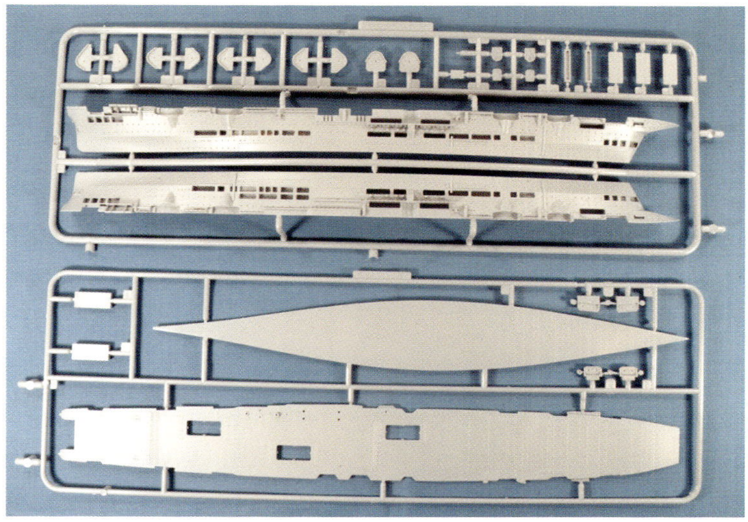

pompom should be omitted. The protecting ramps around the lifts were removed in April 1939.

***Ark Royal* Final Battle:** This variant has some changes from the 1941 version. The instructions have been updated to include a camouflage scheme and the revised bridge with the plated-in director platform; the forward windbreaks are deleted. The decal sheet reverts to the one included in the 1941 kit.

Only one of the aircraft trees with the Fulmar, Skua, Walrus, and folded Swordfish is included, along with the single tree of the older Aoshima Royal Navy Plane set. There are two new parts trees containing two Albacores on each – one with folded wings and the other with wings spread. The ship carried Albacores for a brief period in October 1941 when ferrying them to Malta. The additional tree of boats is also present, and the revised bridge parts are on a new tree replacing the one that had the earlier open bridge piece.

Also included is a photo-etch sheet containing the degaussing cable with an instruction sheet of photos and a diagram showing where each piece is fitted, although the layout is not correct. Aoshima has done a commendable job providing the revised bridge and the degaussing cable. For strict accuracy the protecting ramps around the lifts should be removed and splinter shields added to S1, S2, and S4 pompoms. The ship did not carry Skuas at the time of her loss.

There is also a tree of 21 parts for a full-hull Type VIIC U-boat.

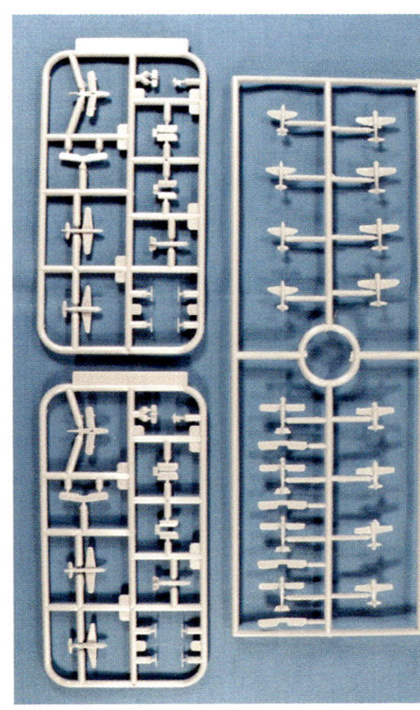

Review models courtesy of **Neil Ramage**

## AOSHIMA *HERMES* 1:700 Scale

This kit depicts *Hermes* in her April 1942 configuration. The instructions are on a folded double-sided sheet with profile and overhead views of the camouflage schemes and decal placement for the ship and aircraft; a drawing of all the kit components is included. Paint references are given for the Mr Color and Aqueous Hobby Color paint ranges with translations in various languages. The included decal

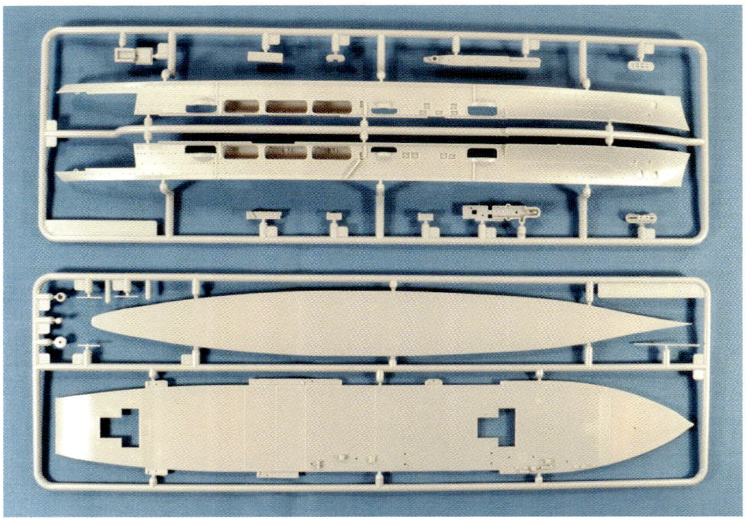

sheet has flight deck and aircraft markings.

Four Swordfish from the Aoshima carrier-borne Set No 1 are included and there are three other Swordfish on the kit's parts trees, one with folded wings. These extra aircraft are of the same level of detail, but the open cockpits are not present.

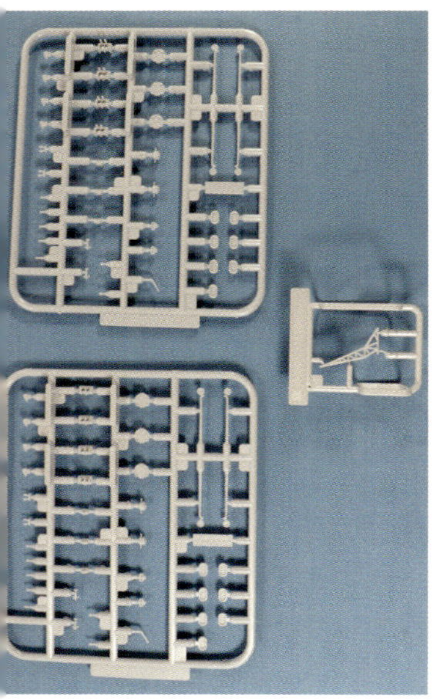

The quality of the moulding is excellent with well-defined smaller parts and many fine details. The hull features a clearly defined bulge along the water line, recessed portholes, plating strakes, mesh over intake vents, sponsons for the 5.5in guns, and cut-outs for the anchors. The island comprises 19 pieces featuring portholes, piping, ladders, hatches and the 500-gallon fuel tank. The flight deck has various raised lockers, arrester wires, and the forward windbreak. The very finely moulded crane has open lattice work, the lattice for the funnel cap is moulded in place, and the roller door to the hangar from the quarter-deck is well defined.

There are few issues with this excellent kit. The boats have internal thwarts and are generic types. The island petrol tank is only semi-rounded due to it being moulded in place. For this time period, the tub at the front of the control top should be present along with splinter shields outboard of the 4in guns. The 20mm mounts are purely speculative.

***Hermes* Battle of Ceylon Sea:** This kit is also the 1942 configuration with several changes and the same issues. The Swordfish from Set No 1 have been replaced with one tree of the older Aoshima Royal Navy Plane set. There are three trees of Japanese aircraft – each contains four Vals, four Kates, and four Zeroes. Only the Vals are used in this kit with metal rods and bases included to display them in flight. The decal sheet includes Japanese aircraft markings. A small etch sheet is also included containing the crane, landing gear and struts for the Swordfish, and aircraft propellers. It is interesting to note that the box art shows *Hermes* with a wooden flight deck.

***Hermes* *Richelieu* Attack:** This kit is *Hermes* in her early war configuration without the homing beacon above the control top or splinter shields for the 4in guns. The flight deck markings have been modified and there is no camouflage scheme. The Swordfish from Set No 1 have been replaced with one tree of the older Aoshima Royal Navy Plane set. A very interesting addition is an 'Insect' class gunboat. At this time period, the tub at the front of the control top was fitted and there were certainly no 20mm guns.

## AOSHIMA ***ILLUSTRIOUS/VICTORIOUS*** 1:700 Scale

These were the first injection moulded kits of the any of the *Illustrious* class in the popular 1:700 scale. The instructions are on a single double-sided sheet with comprehensive historical notes and painting instructions on one side. No decals are included.

The single-piece waterline hull has all of the hull cut-outs and inboard bulkheads moulded in place with an opening for the quarterdeck; there is a separate base plate. The air wing consists of one tree of the older Aoshima Royal Navy Plane set. Moulded-on detail is sparse on both hull and flight deck. At the time of release, Aoshima was the poorest of the Japanese companies in terms of detail and fit and this kit is a reflection of that. Most of the hull openings are very shallow and shaped incorrectly, the placement of some is incorrect, and most of the inboard bulkheads

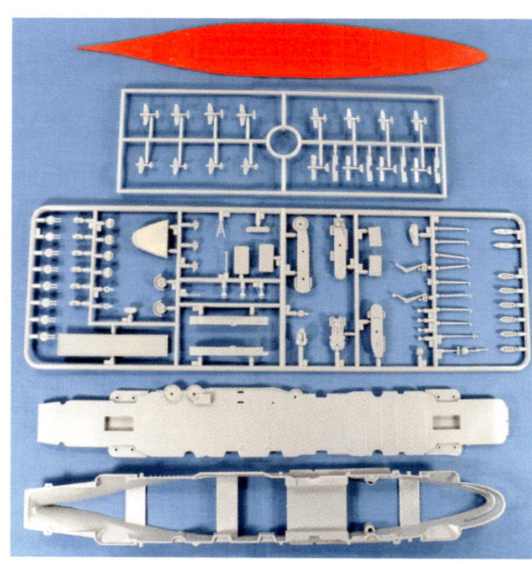

have roller doors engraved into them. There are no hull openings under the bow, no armour belt, the anchors are too low on the hull, and the boat galleries are connected by a sloping section which is inaccurate. The quality of the kit parts is poor with very little detail, the pompoms and main directors are poorly shaped, the boats are crude, and the piece for the open quarterdeck opening protrudes from the hull side when fitted.

This kit would be best built straight out of the box; those desiring accuracy and detail should direct their efforts to the new tooled version.

***Victorious***: This is exactly the same kit with the addition of twelve 20mm single guns and three galleries to place them on. The placement of the galleries on the hull is very vague and they do not appear on subsequent assembly step illustrations.

## AOSHIMA *ILLUSTRIOUS*/*VICTORIOUS* NEW TOOLING

1:700 Scale

This kit depicts *Illustrious* as she appeared up until her bomb damage in January 1941. The instructions are on a folded double-sided sheet with profile and overhead views for three different paint schemes covering early 1940 to mid-1942. Decal placement, paint schemes for the aircraft, and a drawing of all the kit components are also included. Paint references are given for the Mr Color and Aqueous Hobby Color paint ranges with translations in various languages. The included decal sheet has flight deck and aircraft markings.

The waterline hull is split into port and starboard halves. The base plate fits inside the hull, eliminating any seam along the waterline. The air group is made up of two trees of the Aoshima carrier-borne Set No 1 and one tree of Set 2. A tree featuring USN aircraft with two SBD, two TBD, and four Wildcats is also included, but only the Wildcats are relevant. Moulded-on detail is very good, and the smaller pieces are well moulded and feature fine detail.

Aoshima has come a long way from their earliest releases and this is a very fine kit with just a few minor issues: the boats are generic types, the air group is small with different types from different time periods and the 1942 camouflage scheme is not applicable as the changes made after January 1941 are not reflected in the kit.

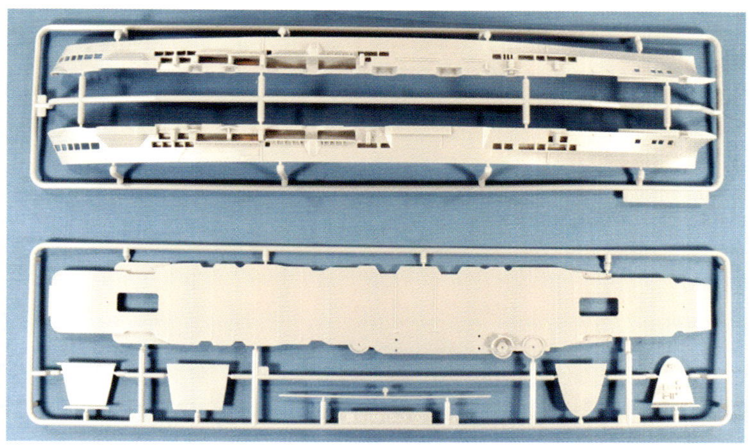

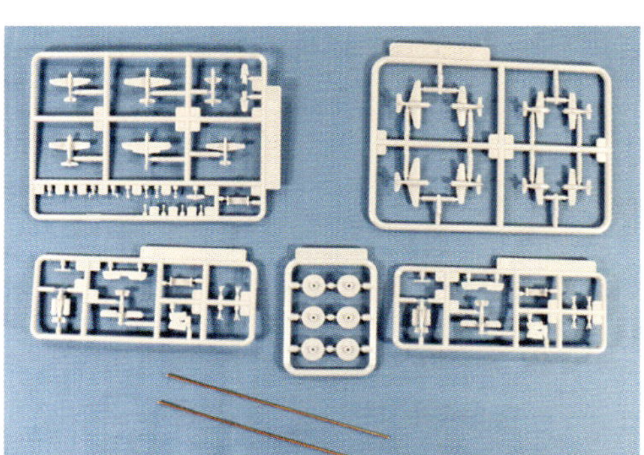

***Victorious***: This is the same kit as *Illustrious* with the following minor differences. There are two camouflage patterns shown: May 1941 and mid-1941 to mid-1942. The flight deck markings on the decal sheet have changed, two parts trees with alternate antenna masts are omitted, the flight deck radar mast has two platforms, and the wind deflector at the front of the island is present.

***Illustrious* Benghazi Attack:** This variant of the *Illustrious* kit differs modestly from the base kit. The boats are specific types on two new parts trees featuring enhanced detail; the instructions have been modified accordingly. The air group has been changed to four trees of the Aoshima carrier-borne Set No1, and two trees each containing a single Fulmar, Skua, Walrus, and folded Swordfish as found in the *Ark Royal* kit. There are also two trees with three Italian SM79 torpedo aircraft on each.

Review models courtesy of **Brian Joslin**

## FLYHAWK *ILLUSTRIOUS* 1940/*FORMIDABLE* 1941

1:700 Scale

Packed with parts and detail, this kit depicts *Illustrious* as she appeared from late 1940 up to her May 1941 refit at Norfolk, Virginia. The two double-sided instruction sheets are colour coded with full colour profile and overhead views of the ship and aircraft showing camouflage paint schemes and decal placement with references for the Mr Hobby, Tamiya and WEM Colourcoats paint ranges. The decal sheet has flight deck and aircraft markings. A photo-etch sheet containing aircraft parts and the safety barriers is also included.

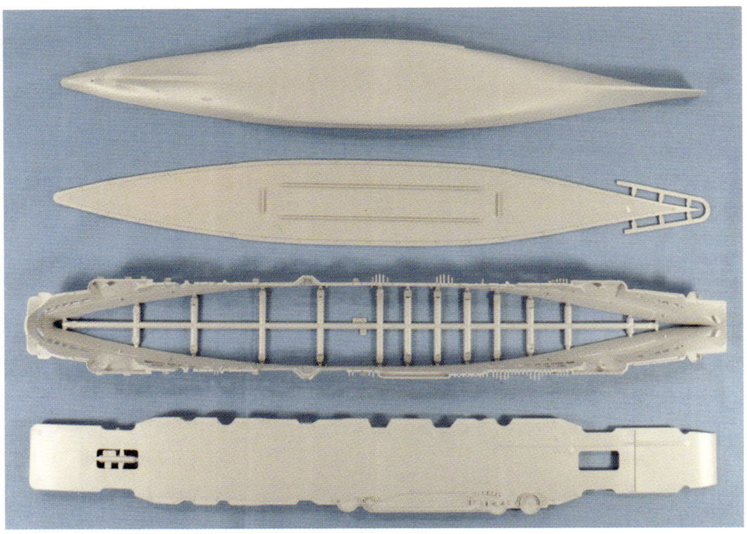

The single-piece waterline hull has cut-outs for all the hull openings with a choice of base plate or full underwater hull. The single-piece flight deck has cut-outs for the lifts. The superstructure, weapons, and smaller fittings are on 23 parts trees. There is a single-piece hangar deck that fits snugly inside the hull and the single-piece inboard bulkheads fit atop that, making a continuous bulkhead that runs the length of the ship behind the hull openings. The moulding quality is superb on all the parts, both large and small. The air group is eight Swordfish and six Fulmars from the Flyhawk WW2 Royal Navy Aircraft Set 1.

This is another superbly researched, engineered, and crafted kit from Flyhawk, but there are no stands included for a full-hull version.

***Formidable* 1941:** This variant of the kit has a few differences. The flight deck is a new part with the correctly shaped raised catapult casing. The camouflage scheme is the one worn by the ship off Crete in May 1941. The etch set has Type 282 and Type 285 arrays for the directors, and a flight deck windscreen. Pompom directors, a Type 279 radar mast, and the windscreen at the front of the island have been added. The air group has been changed to two Swordfish and six Fulmars from Flyhawk Set 1, and three trees with six Albacores. A minor issue is that the ship was not fitted with Type 282 or 285 radars in May 1941.

Review models courtesy of **Flyhawk Models**

## FLYHAWK *HERMES* 1942

1:700 Scale

Depicting *Hermes* in her final appearance, this kit is packed with detail. The double-sided instruction sheet is colour coded and there are full colour profile and overhead views of the ship and aircraft showing paint schemes and decal placement with references for the Mr Hobby, Tamiya and WEM Colourcoats paint ranges. The decal sheet has flight deck and aircraft markings, and an information card with box art is included. A very comprehensive photo-etch sheet is also included with all the flight deck and superstructure railings, accommodation ladders, mesh platforms, and grilles needed for the kit. The single-piece waterline hull has cut-outs for all the hull openings and there is a choice of a base plate or full underwater hull. There

are 17 parts trees for the superstructure and a myriad of smaller parts. The air group consists of eight Swordfish from the Flyhawk WW2 Royal Navy Aircraft Set 1.

The quality of the mouldings is superb everywhere; all the parts both large and small are highly detailed and have a very delicate look. The flight deck is fully detailed, even on the underside; in fact the entire hangar is fully detailed which is quite remarkable as none of it will be visible should the modeller choose to leave the lifts in the up position. The kit is magnificent, superbly engineered and crafted; all the parts fit together extremely well.

There a few problem areas. The waterline bulge is depicted as an armour belt, careful sanding will be required to remove the forward and after perpendicular edges. There should be splinter shields outboard of the 4in guns. The light AA is incorrect: the ship was never fitted with a quad pompom, a 0.5in mount should be fitted instead in that position, and the 20mm mounts are speculative.

***Hermes* 1937 Coronation Fleet Review:** There are a few changes for this version of the kit. The air group has four Swordfish from the Flyhawk Set 1 and two trees each containing a single Walrus; the Walrus is as highly detailed as the Swordfish and with the option for spread or folded wings. The decal sheet has the 'HR' identifying letters and a large selection of signal flags for dressing the ship overall.

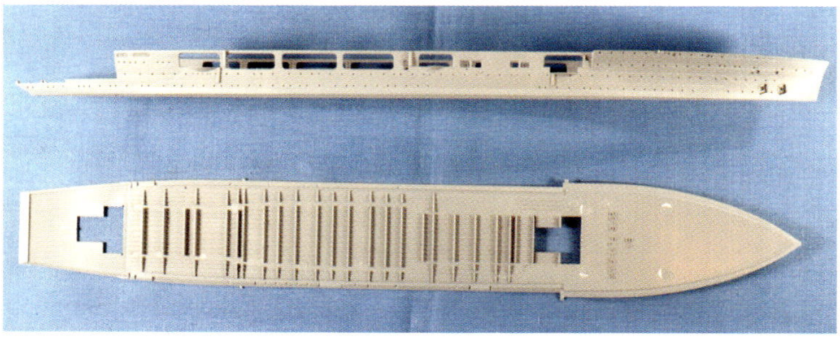

There is a new piece for the lower island; the island is fitted with saluting guns and does not have the tub at the after end, and the homing beacon is not fitted. Life rafts are placed along the edges of the flight deck. A 0.5in mount should be placed forward of the 4in guns ahead of the island.

## FORCES OF VALOR *ARK ROYAL* 1:700 Scale

This unique item is a pre-assembled and pre-painted die cast model, not an actual kit. Packaged in a painted seascape, it is actually full hull with a display stand. The only parts not attached are the propellers and shafts, the display stand, and the aircraft, all of which are included along with a tube of glue. For those not too interested in complete accuracy, the model is a decent replica overall with lots of hull, flight deck, and island detail. The air group comprises five Fulmars, four folded Swordfish, and six Swordfish with wings spread. All single pieces, they are very crude and would be best replaced with aftermarket versions.

■ Review model courtesy of **Imperial Hobbies**

## GAGA *IMPLACABLE* 1:700 Scale

Depicting *Implacable*'s wartime appearance, this is one of the last of the WWII carriers to finally be produced in kit form. Its appearance provides an interesting glimpse into the future of the hobby as it is produced entirely by 3D printing and is only available on an online auction site. The instructions consist of five drawings on a single sheet (featuring a non-existent lower hull) showing placement of the smaller parts. There is a colour coded version available on the auction site where the model is sold.

The single-piece waterline hull is quite remarkable as it has all the structural pieces in place. The quarterdeck and forecastle openings are quite phenomenal as there are no bulkheads behind them; they are hollow inside with detailed decks. There are some ninety 3D printed parts and the quality is uniformly excellent, with much fine detail including lattice work on the masts and crane booms; the boats with cabins are hollow. Not much clean-up will be required, and there is no casting block to be removed! No aircraft are included.

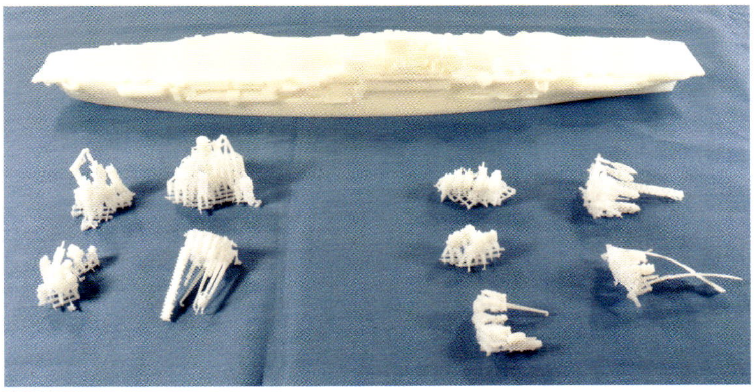

The overall shape of the hull, island, and flight deck is very accurate, but there are some problems for the modeller to overcome. 3D printing often leaves striations as part of the process and these are present on

all the decks and the hull. They can be sanded out easily from the flight deck, but getting them out of the smaller decks will require some finesse. There are no portholes, the stern is shaped incorrectly, the flight deck is devoid of detail and the after tripod mast is not included.

The instructions are rather vague as to the placement of the smaller parts; the colour coded instructions are only slightly more helpful. Correcting the stern shape will not be too difficult, adding the missing detail, especially on the flight deck, will require some careful measuring and scratch-building. Due to the nature of the printing process which has all the structures and hull openings already in place, converting to *Indefatigable* will require major scratch-building.

Review model courtesy of **Richard Harden – Tom's Modelworks**

## TRUMPETER *ARK ROYAL* 1939 — 1:700 Scale

Meant to represent *Ark Royal* in her early war fit, this kit is packed with parts. The instructions are in booklet form with drawings of all the kit components, and there is a separate full colour painting and markings drawing for the ship and aircraft with references to the Mr Hobby, Vallejo, Model Master, Tamiya, and Humbrol ranges. The decal sheet has flight deck and aircraft markings and there is an etch sheet with the crane booms, antenna masts, funnel grille, and Type 285 radar.

The waterline hull is split into port and starboard halves with a moulded-in base plate, so no troublesome sanding along the waterline. The forecastle deck and quarterdeck pieces have separate bulkhead parts. Moulded-on hull detail is extensive and the quality of the kit parts is excellent throughout, with many fine and delicate details. The air group comprises five Swordfish, four Fulmars, and four Skuas, all composed of superbly detailed multiple parts, although there is no folding wing option for any of the aircraft.

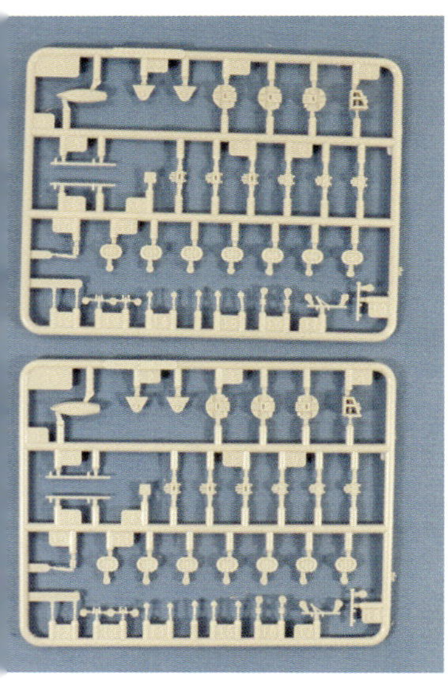

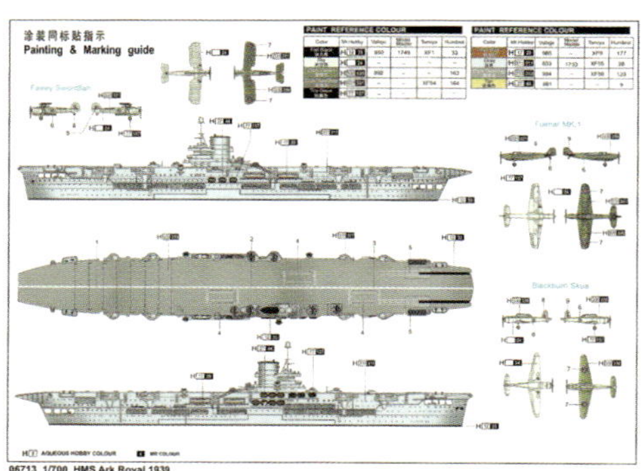


To be strictly accurate for 1939, the degaussing cable, splinter shielding for the forward pompoms, and the port side pompoms should be omitted. The splinter shields are the wrong shape in any case and the placement of the degaussing cable is not accurate. The ship did not carry Fulmars at that time, and was never fitted with Type 285 radar. The fore and aft hull knuckles are not well defined but otherwise there is very little to quibble over with this excellent kit and the issues can be easily addressed. S2 pompom should be higher than S1; the pompom directors are shaped like searchlights; the swing-out catwalk on the port side of the island is not present. There should be a pedestal under the range finder platform ahead of the bridge and there is no forward windbreak or safety barrier.

## AIRFIX *ARK ROYAL* — 1:600 Scale

Now part of the Vintage Classics range, this kit made its first appearance in 1966. The single-page foldout instructions have a simple painting guide for the ship with references to the Humbrol range.

The full hull is split into port and starboard halves with cut-outs for all the hull openings, while the single-piece flight deck has cut-outs for the lift platforms. The island is made up of eight pieces but does not match either configuration as it lacks the forward rangefinder platform. The air group consists of six single-piece Fulmars, and six three-piece Swordfish; two of each type have folded wings, but there is no detail on any of the aircraft. Moulded-on hull detail is limited but the flight deck features catapults, safety barrier, and arrester wires. There is little detail on the kit parts and many of them, particularly the boats, have casting dimples that will need to be filled.

There are some areas where improvements can be made. There is no armour belt or degaussing cable and no forecastle opening under the bow. The forward windbreaks and catapult loading-on platforms are not present; both forward pompoms are at the same level, there is no splinter shielding for the flight deck pompoms, and the safety barrier is a solid protruding rib. There is no funnel cap, swing-out catwalk, or forward rangefinder platform.

These issues and the general lack of detail are a reflection of the kit's age and can be tidied up with aftermarket items and some scratch building. At over 40cm in length it makes an impressive finished model.

## AIRFIX *VICTORIOUS* — 1:600 Scale

One of the first Airfix ship kits, it made its first appearance in 1959 when the ship had just recommissioned after her modernisation. Long out of production after mid-1980, it was re-introduced to the range in 2019. Instructions are in booklet form and include drawings of the ship and aircraft with paint references to the Humbrol range. There is a small decal sheet containing the pennant number and minimal flight deck markings.

This is a full hull model, split into port and starboard halves with cut-outs for all the hull openings. On the single-piece flight deck there is a cut-out for the after lift platform. The air group are all single-piece aircraft: six Sea Vixens, four Scimitars, and four Skyraiders (one with folded wings). The shapes are adequate, but there is little detail on any of them and the undercarriages are simple nibs. Moulded-on hull detail is limited and the kit parts have very little detail with the boats simplified generic

types. Considering its age, the moulds are still in good shape and there are very few ejection pin markings. The basic hull and flight deck shapes are good and there is a lot of scope for super-detailing.

Review model courtesy of **Ernie O'Brien**

## HELLER *ILLUSTRIOUS* — 1:400 Scale

First released in 1979 and re-issued as recently as 2012, this kit depicts *Illustrious* in her pre-May 1941 appearance. The instructions are on a large size double-sided sheet with another sheet having profile and plan drawings of the ship and aircraft; camouflage schemes are shown for the aircraft and for the ship in her mid-1942 to June 1943 scheme. Paint references are given for the Heller range. The decal sheet has flight deck markings and roundels for the aircraft.

The kit components are moulded in two colours, grey and green, with the full hull split into port and starboard halves with cut-outs for all the hull openings. Smaller parts are on six trees; there is a tree of injection moulded railings as well and there is even an attempt at flight deck safety netting. Moulded-on detail is reasonable and the smaller parts are sharply defined, but lack detail as do the inboard bulkheads and quarterdeck pieces. The air group comprises six Swordfish and six Martlets.

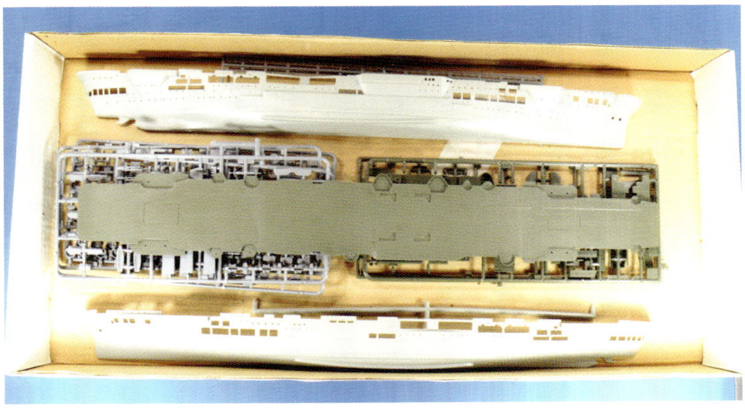

The Martlets can be assembled with wings folded or spread; the Swordfish only have spread wings. There is little detail on either type.

There are a few problems with accuracy. The after lift is wider than the forward lift. Careful attention needs to be paid to the hull openings, as many are not correctly

shaped or placed. This also applies to the flight deck cut-outs. The sheer size of all the components makes this a very impressive kit with plenty of scope for super-detailing. For the pre-May 1941 time period, the wind deflector at the front of the island and the Type 285 radar arrays should be left off. The Martlets would not have been embarked at that time. For the time period of the 1942–1943 camouflage scheme, the modeller will have to add additional light AA galleries and flatten out the after round down.

Review model courtesy of **Neil Ramage**

## MERIT *ARK ROYAL* 1939

1:350 Scale

This is a scaled up full hull version of the Trumpeter *Ark Royal* 1939 1:700 scale kit, the larger scale allowing much more detail. There are sixteen pages of instructions in booklet form with drawings of all the kit components, and a separate full colour painting and markings drawing for the ship and aircraft with references to the Mr Hobby, Vallejo, Model Master, Tamiya, and Humbrol ranges. The decal sheet has flight deck and aircraft markings and there are eight photo-etch sheets with railings, antenna masts, pompom details, windbreaks, Type 285 radar, funnel grille, safety barrier, and flight deck netting.

The hull is split into port and starboard halves with cut-outs for all the hull openings, with the flight deck a single piece. There are twelve trees of parts, adding a host of high-quality detail to the hull. The island features the early bridge with the lower rangefinder platform and is made up of twelve main pieces with plenty of photo-etch ladders, supports, and railings. Moulded-on hull and flight deck detail is the same as the Trumpeter kit but the fore and aft hull knuckles are much crisper. The air group is made up from the Trumpeter 350 scale aircraft sets with five Swordfish, four Fulmars, and four Skuas.

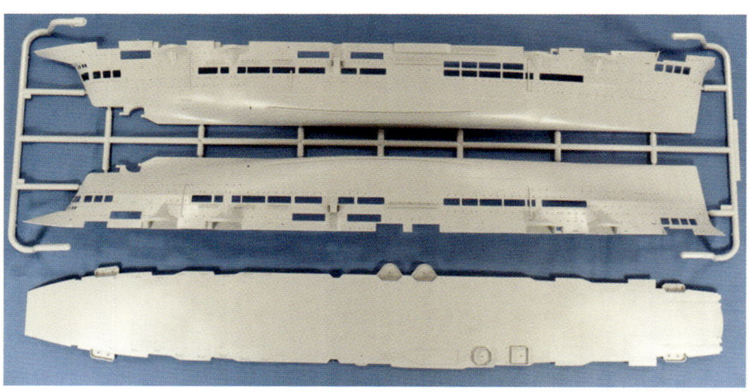

For the 1939 time frame, this kit has the same inaccuracies as the Trumpeter kit regarding the degaussing cable, pompom splinter shielding, the port side pompoms, Fulmars, and Type 285 radar. The pompom directors, swing-out catwalk, range finder platform pedestal, and ribbing on the internal bulkheads are also common problem areas. In addition, the waterline is not marked and will require some careful measuring to paint the underwater hull. The anchors are very undersized. Given the size of the kit, it is a pity that the manufacturer did not provide separate lift platforms and lift wells; providing them will require some careful scratch building. This is still a very impressive kit and will make a very good model right out of the box.

Review model courtesy of **Neil Ramage**

While not the subject of a detailed review, the Orange Hobby kit of HMS *Victorious* as reconstructed can be seen in the Modelmakers' Showcase section.

# ACCESSORIES PHOTO-ETCH

## FLYHAWK *ILLUSTRIOUS* 1940 ETCH/DETAIL SET

1:700 Scale

Made for their *Illustrious* 1940 kit, this comprehensive set has five etch sheets, a three-page full colour instruction sheet, and a photo of the completed model with the detail parts in place. Also included is a set of metal parts: 4.5in barrels, foremast, and rod for cutting into smaller pieces for yardarms and catwalk supports. The quality of the etched parts is excellent with a lot of relief etching. The instructions are thorough, but the diagrams are small and require careful study to determine where the parts are to be placed. This entire set is included in the Special Edition of the *Illustrious* 1940 kit.

■ Review set courtesy of **Flyhawk Models**

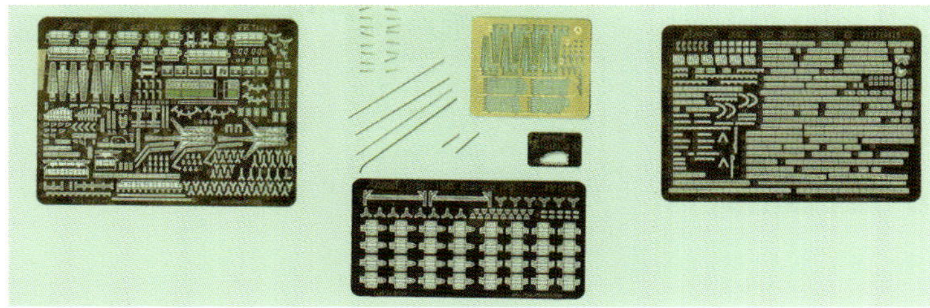

## TOM'S MODELWORKS ***ILLUSTRIOUS*** CLASS ETCH SET

1:700 Scale

Designed specifically for the new-tooled Aoshima *Illustrious* and *Victorious* kits, this etch sheet is packaged with Tom's WWII British Radar etch sheet. There are eight pages of instructions with drawings, notes, and photos giving complete details for using the etch sheets. Any WWII configuration can be modelled as there are parts for all relevant weapons and radar, including the early and late war foremast, and the late war mainmast carried by both *Victorious* and *Formidable*. This is an extremely well thought out and produced set; the late war masts and the stern AA gallery make unique additions to the basic kits.

■ Review set courtesy of **Richard Harden – Tom's Modelworks**

## TOM'S MODELWORKS ***HERMES*** CRANE ETCH SET

1:700 Scale

Meant to replace the plastic flight deck crane on the Aoshima or Flyhawk kits, this is a very simple and effective way to enhance the appearance of a finished model.

■ Review set courtesy of **Richard Harden – Tom's Modelworks**

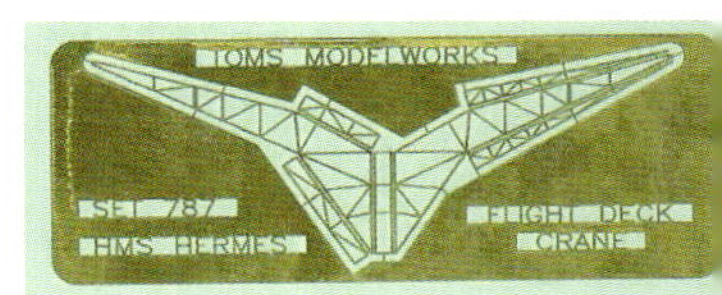


## WHITE ENSIGN ***ARK ROYAL*** ETCH SET

1:700 Scale

This single sheet set is designed for the Aoshima *Ark Royal* kits and can also be used on the Trumpeter kit. Five pages of instructions and diagrams give precise fitting, folding, and placement details, along with notes for modifying kit parts. A parts list is included along with a diagram of the etch sheet to aid in identifying specific pieces so modellers should experience little difficulty assembling and placing the parts. This is a very practical and comprehensive set, covering weapons, cranes, antenna masts and flight deck early war safety barrier, catwalks, arrester wire guides, and wind deflectors. Aircraft details include landing gear, propellers, and Swordfish wing struts.

■ Review set courtesy of **Richard Harden – Tom's Modelworks**

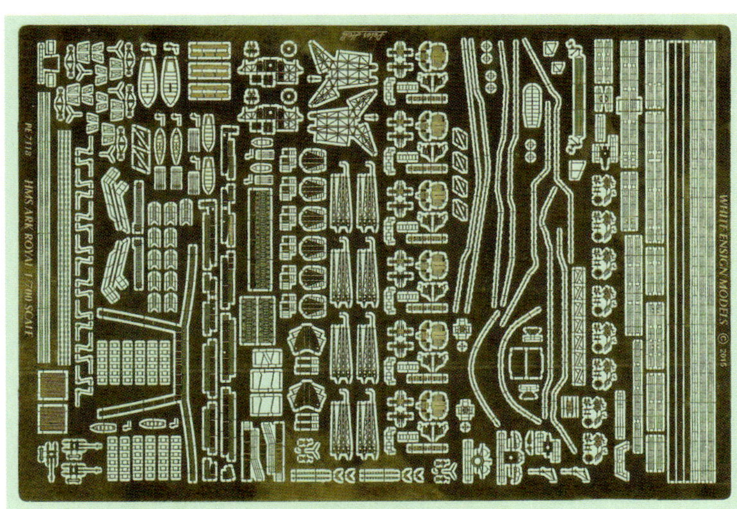

## WHITE ENSIGN ***ARK ROYAL*** ETCH SET

1:720 Scale

Designed specifically for the Revell 720 kit, this set can also be used for any of the 700 scale *Ark Royal* kits. Two pages of instructions and a parts list with diagram of the sheet are included. It has many features in common with the corresponding 700 scale set but there is no degaussing cable, elevator wind deflectors, life buoys, or specific railing sections for the hull openings, and there are no supports for the multi-part stern catwalk. Additions in this set include tubs for S1 and S4 pompoms, a platform for S2 pompom, screens between S1 and S2 and between S3 and S4, catapult shuttles, hatches, and accommodation ladders.

■ Review set courtesy of **Richard Harden – Tom's Modelworks**

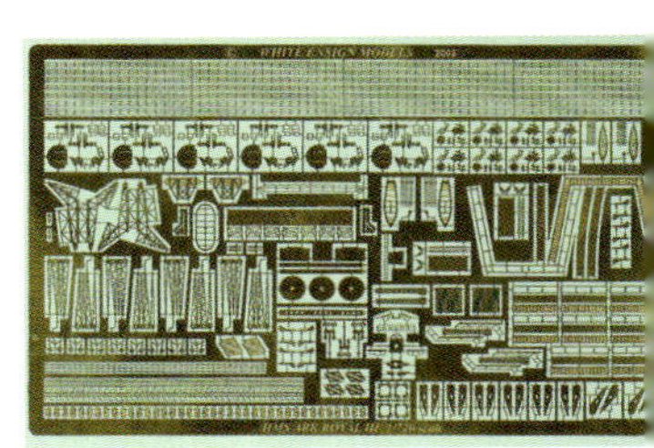

## WHITE ENSIGN *ARK ROYAL* ETCH SET 1:600 Scale

Designed for the Airfix kit, this set has a single etch sheet and two pages of instructions, a drawing of the parts and a complete parts list. Many of the features are common with the corresponding 700 scale set but there are no flight deck catwalks, elevator wind deflectors, life buoys, specific railing sections for the hull openings, or supports for the multi-part stern catwalk. Additions in this set include tubs for S1 and S4 pompoms, a platform for S2 pompom, screens between S1 and S2 and between S3 and S4, catapult shuttles, hatches, platform supports, davits, and accommodation ladders. Unique to this set are rails to be fitted inside the catapult tracks, a screen for the batsman, and lift surrounds to detail the sides of the lift wells. Another very good and practical set.

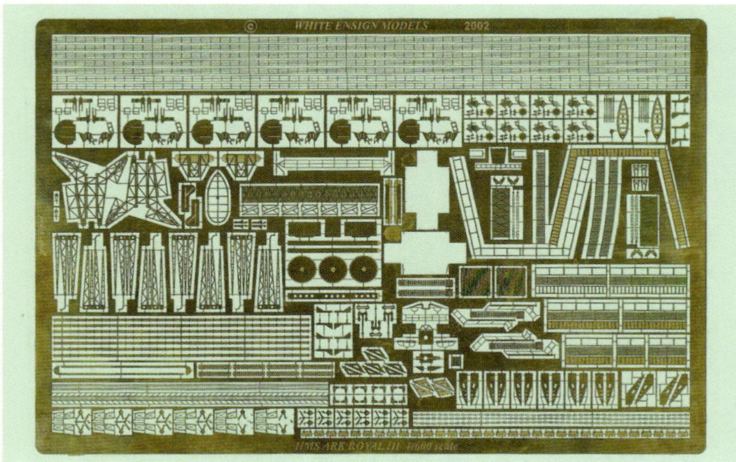


Review set courtesy of **Kerry Jang**

## WHITE ENSIGN *ILLUSTRIOUS*/*VICTORIOUS* ETCH SET 1:400 Scale

This set of two etch sheets is designed for the Heller *Illustrious* kit. Seven pages of instructions with diagrams and notes make the construction and placement of the various elements very clear. There are also drawings of the etch sheets and a full parts list.

Apart from the usual generic items, parts for building the early and late war foremast, the after tripod mast fitted to *Victorious*, stern 20mm gallery, YE homing beacon, and twin and single 20mm mounts allow any WWII time period to be modelled. This is a very high quality and well-presented set. The large size of the many pieces will make assembly and placing very straightforward.

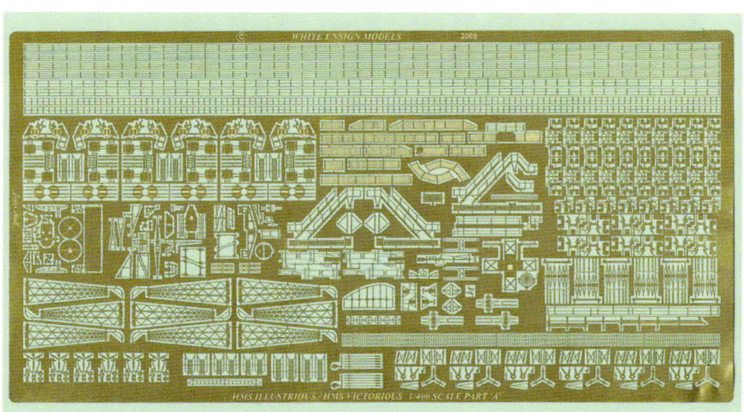


Review set courtesy of **Richard Harden – Tom's Modelworks**

## WHITE ENSIGN *ARK ROYAL* ETCH SET 1:350 Scale

Designed specifically for the Merit kit, this two-sheet etch set has eight pages of instructions, a parts list, and drawings of both etch sheets.

There are many features common with the corresponding 700 scale set but there are no catapult loading on platforms, safety barrier, arrester wire guides, anchors, generic ladders or aircraft parts as those details are supplied with the kit. On the other hand, some fittings, like the catwalks under the side windbreaks, are more complex, and there are a number of parts specific to this set, including the swing-out catwalk on the island.

A very impressive set, the large size of the parts allows for more fine detail with a lot of relief etching. Including the parts to convert the boats is a particularly inspired choice.

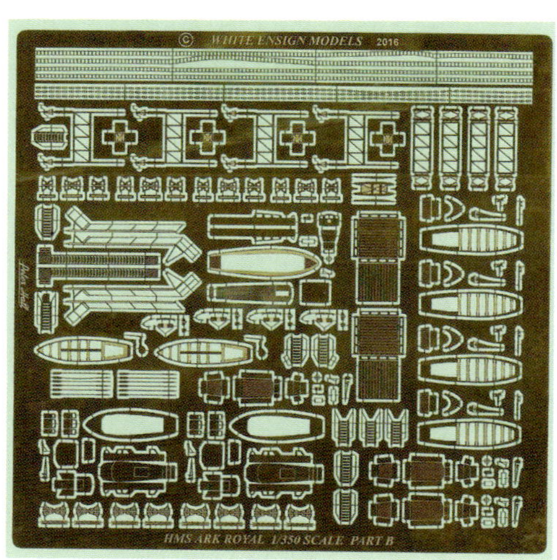


Review set courtesy of **Richard Harden – Tom's Modelworks**

# FITTINGS

## BLACK CAT 3D PRINTED PARTS 1:350 Scale

Black Cat Models have a large range of 350 scale 3D printed parts for various navies. All feature high quality, accurate, and precise detail; little cleanup is required once the parts are removed from their printing platforms.

For the RN carriers, Black Cat has two sets of boats (35ft Fast Motor Boat and 45ft Motor Launch) and one set of Carley floats. A unique offering is a set of 55 flight deck crew, flight crew, and DLCO figures in multiple poses – and unlike the photo-etch equivalents, these figures have depth and definition with clearly defined clothing. Black Cat also makes three different sets of highly detailed light AA weapons: single 20mm, octuple 2pdr pompoms MK V, and quad 0.5in Mk I.

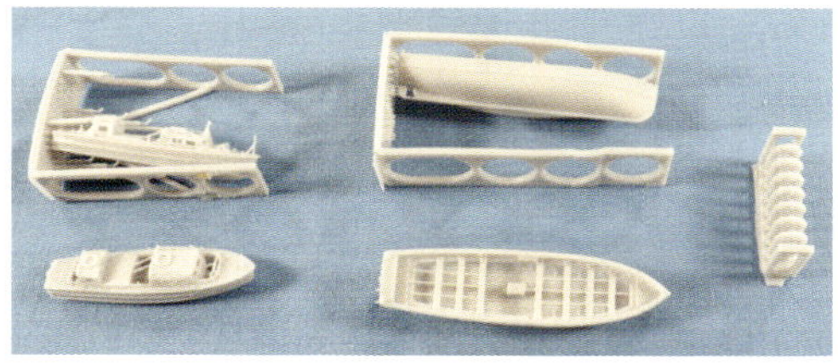

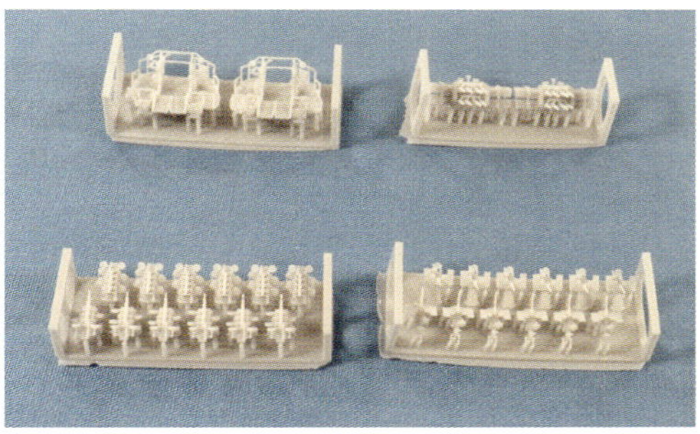

■ Review parts courtesy of **Black Cat Models**

## L'ARSENAL QUAD 2PDR POMPOMS 1:700 Scale

These 3D printed pompoms have deck and gun mount pieces. The deck features tread plate and railings, the gun mountings have the ammunition holders and clearly defined separate barrels. Six are provided per set.

■ Review pompoms courtesy of **L'Arsenal Models**

## MICRO MASTER 3D PRINTED REPLACEMENT PARTS 1:700/600/350 Scales

Micro Master Models has an extensive range of 3D printed replacement parts for Royal Navy ships. Well researched, they are resin printed and available in many scales; the website contains technical information for many of the parts. All parts are of the highest quality and feature accurate, precise, and very delicate detail; very little cleanup will be required once the parts are removed from their printing platforms. Where appropriate, the fittings and weapons have locating lugs printed in place. This review concentrates on items appropriate for RN Carriers.

Boats suitable for RN carriers include 35ft Fast Motor Boats, 25ft Fast Motor Boats, 32ft Cutters, 32ft Motor Cutters, 27ft Whalers, and Carley floats. Fittings for RN carriers include pompom directors, W/T doors, searchlights and signalling projectors, HACS directors, and anchors. Of particular note, the HACS MkIV director is impressively detailed with a rangefinder, crew positions, rungs, telescope, and raised detail inside and outside of the mounting. A second version has a precisely detailed Type 285 radar array attached.

For the 1:700 scale carriers, weapons include single 4in/45 guns suitable for HMS *Hermes* and twin 4.5in BD mounts for the *Illustrious* class, as well as a range of single and twin 20mm on pedestal mounts and the twin 20mm MkV mounting. Quad 0.5in and highly detailed octuple pompoms are available in 1:700/600/350 scales.

Needless to say, the detail at 1:350 scale is particularly striking, but even at 1:700 these 3D printed pieces are a vast improvement on anything likely to be provided in even the best kits.

■ Review parts courtesy of **Simon Percival and Jacqueline Franco at Micro Master**

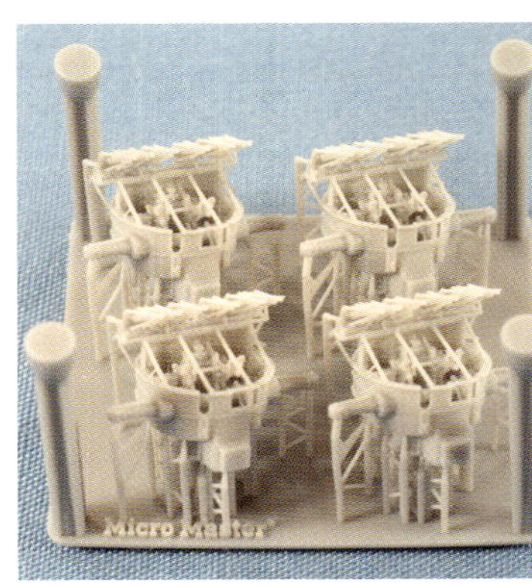

# CARRIER AIRCRAFT

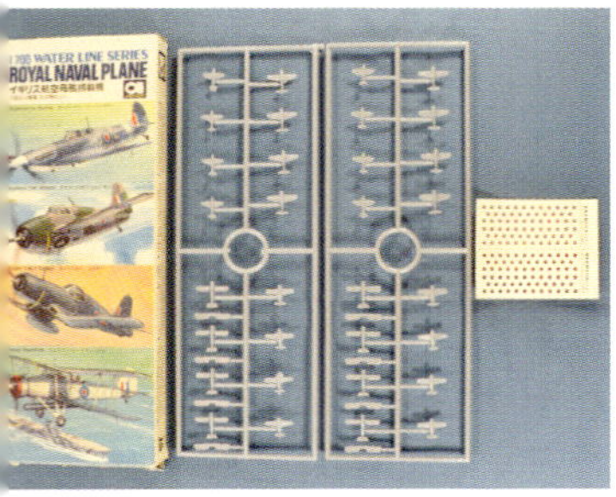


## AOSHIMA ROYAL NAVY PLANE 1:700 Scale

This set contains Corsair, Martlet (Wildcat), Seafire, and Swordfish aircraft, four of each type on two trees giving a total of eight of each aircraft in the set. Two types of roundels are on the enclosed decal sheet. Profile and overhead drawings showing typical camouflage schemes are on the back of the box. The aircraft types are recognisable but detail is limited to open cockpits on the Swordfish and cockpit canopies on the rest. The wings of the Seafires are too short.

This set is of limited value as it has been surpassed by much better products, but it continues to be included in the new tooled Aoshima *Ark Royal* and *Hermes* kits.

## AOSHIMA BRITISH CARRIER-BORNE AIRCRAFT SETS NO 1 & NO 2 1:700 Scale

**Set No 1:** This set of extra aircraft contains Swordfish only. The instructions are on a single double-sided sheet showing decal placement and three different camouflage schemes. Paint references for the Mr Color and Aqueous Hobby Color ranges are given with translations in various languages. The decal sheet contains five different types of wing roundels along with fuselage roundels and fin flashes. Each of the four parts trees contains two Swordfish, one with folded wings, giving a total of eight aircraft. The aircraft are very well detailed with separate landing gear and top wing, torpedo, and open cockpits. There are no propellers or wing struts.

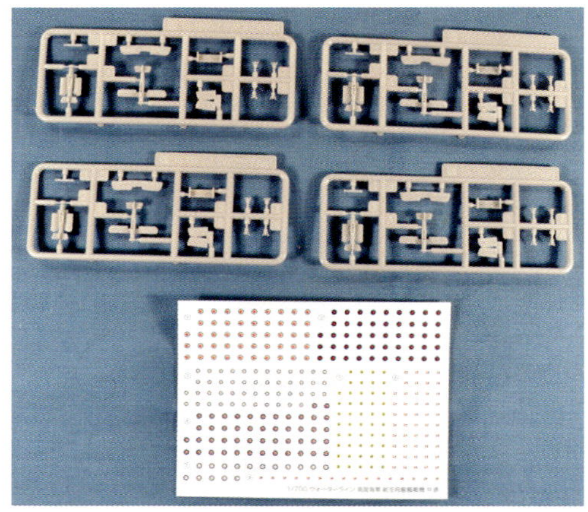

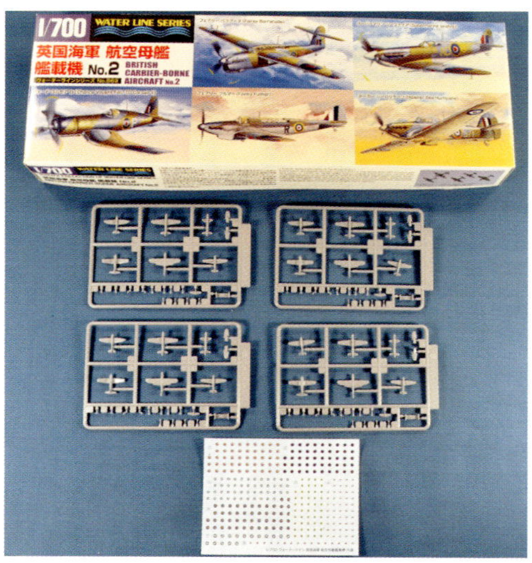


**Set No 2:** This set is the same as the above in terms of instructions, decals, and the level of detail for each aircraft, but there is only one camouflage scheme shown per aircraft. Each of the four parts trees contains two Fulmars (one with folded wings), and a single Seafire, Sea Hurricane, Corsair, and Barracuda, giving a total of 24 aircraft.

Review sets courtesy of **Brian Joslin**

## FLYHAWK WW2 ROYAL NAVY AIRCRAFT SET 1 1:700 Scale

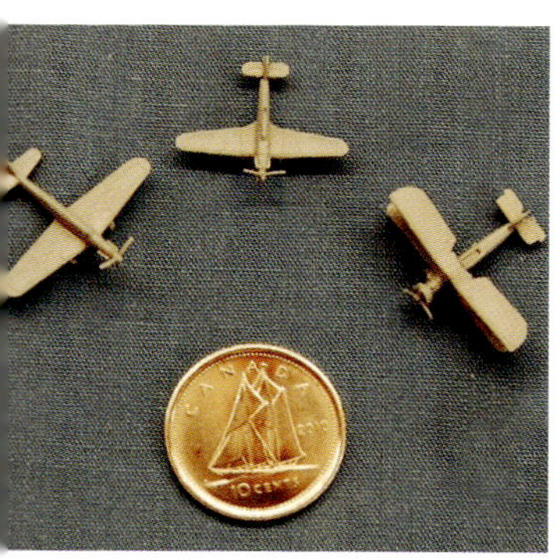

This set contains Swordfish, Fulmars, and Sea Hurricanes, moulded in grey styrene. There is a double-sided instruction sheet and the full colour box bottom contains drawings of each type with painting and decal placement instructions; references are given for the Mr Hobby and Tamiya paint ranges. The decal sheet has wing and fuselage roundels and fin flashes for each type. An etch sheet contains struts and rear seat machine guns for the Swordfish, and alternate propellers for each type.

Each aircraft type is on its own parts tree totalling eight Swordfish, six Fulmars, and four Sea Hurricanes. All the aircraft are remarkably detailed with clearly defined panel lines, elevators, rudders, and flaps. The landing gear and propellers are separate pieces; the landing gear has moulded-in wheels and tyres. The Swordfish and Fulmars can have their wings folded or spread.

Review set courtesy of **Flyhawk Model**

## L'ARSENAL FULMAR/SWORDFISH 1:700 Scale

These two sets are multimedia kits of resin and photo-etch parts. Easy to follow assembly instructions are included. The Fulmar set contains ten aircraft comprising a single resin piece with a detailed canopy, engine cowling, and elevators, and an etch set for details. There is no folded wing option. The Swordfish have a resin fuselage with cockpit and detailed engine, with wings, propeller, landing gear, and wing struts on the etch sheet. The set contains five aircraft, one with folded wings. It will take careful construction, but both kits will result in well detailed aircraft.

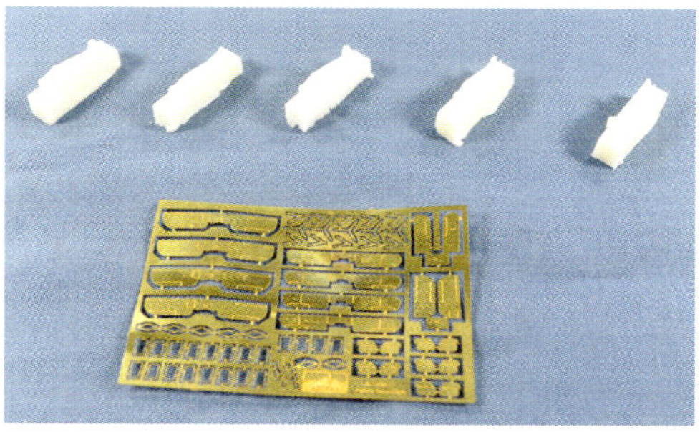

Review aircraft courtesy of **L'Arsenal Models**

## TOM'S MODELWORKS WW2 ROYAL NAVY AIRCRAFT 1:700 Scale

Part of the extensive Tom's Northview Foundry aircraft line, these 3D printed resin aircraft are all single pieces with attached landing gear. Specific RN carrier-based types available include: Albacore, Fulmar, Roc, Skua, Seafire, Sea Gladiator, Sea Hornet, Sea Hurricane, Swordfish, and clipped wing Corsair. Applicable USN types include Hellcat, Wildcat (Martlet), Avenger, and Hellcat night fighter with wing radome. They are well detailed with flaps, elevators, rudders, tail wheels, and clearly defined wheels and tyres on the undercarriages which are accurately shaped for each type. Both folded and spread wing versions are available for appropriate types. Their great appeal is that they are very accurately shaped, require no assembly, and have many unique types allowing specific air groups to be built up easily.

Review aircraft courtesy of **Richard Harden – Tom's Modelworks**

## TRUMPETER FULMAR/SKUA/SWORDFISH 1:350 Scale

Each set features one aircraft type. Moulded in grey styrene, they are packaged six per box with a decal sheet with roundels and fin flashes; the assembly instructions are on the back of the box. The Swordfish comes with an etch sheet containing the wing struts. All aircraft types are extremely well detailed with panel lines, ribbing, elevators, flaps, detailed wheels and tyres on the landing gear, a tail wheel, and rudder. None of the aircraft have a folded wing option.

Review aircraft courtesy of **Neil Ramage**

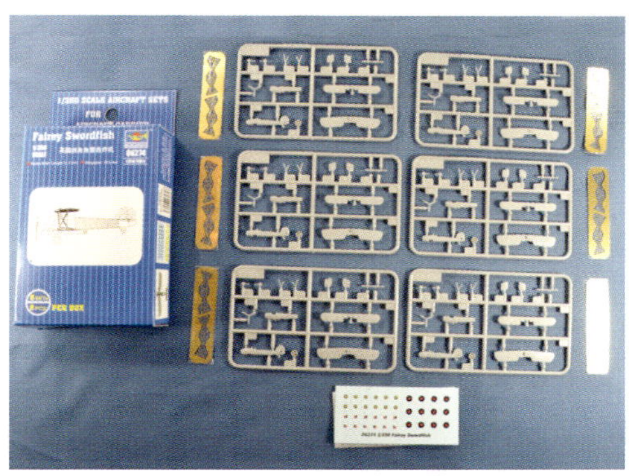

## GOLD MEDAL MODELS AIRCRAFT PARTS ETCH SET 1:700 Scale

This generic sheet can be used to dress up a wide variety of aircraft. It contains non-aircraft specific three- and four-bladed propellers, landing gear, undercarriage doors, and tail hooks.

## WHITE ENSIGN FULMAR/SKUA/SWORDFISH ETCH SETS 1:350 Scale

Designed specifically for the Trumpeter 350 scale aircraft, these sets provide an extra level of detail. Each has a single etch sheet with parts for six aircraft, typically including tail wheel, propeller, wing flaps, undercarriage doors (where applicable), and arrester hook A single-sided instruction sheet with a diagram giving precise folding and fitting instructions is included.

Review sets courtesy of **Richard Harden – Tom's Modelworks**

# Modelmakers' Showcase

## FLYHAWK *HERMES* 1942 1:700 scale By ERICK CHANG

**Erick Chang** writes:

The model represents HMS *Hermes* in early 1942 just before her loss, carrying the Swordfish of 814 Naval Air Squadron. The camouflage scheme was based on information provided by Sovereign Hobbies. The paints used are Sovereign Hobbies Colourcoats; the base coats and flight deck were painted using an airbrush, the camouflage was painted by hand.

The kit parts are very detailed and the fit is excellent. The instructions have to be very carefully followed for the hangar and island as the order of parts placement does matter. The kit's photo-etch parts were used; the aircraft are from L'Arsenal, rigged with 0.05mm metal wire. I used the Flyhawk QF 4in naval gun photo-etch and resin set, the quad pompom is from 3D Model Parts. The flight deck safety netting is scratch built and the model is rigged with stretched sprue, metal wire, and elastic thread.

Total construction time was 11 months. The model won a Gold Medal in its category at the IPMS UK Scale Model World event at Telford in 2018.

# AIRFIX *ARK ROYAL* 1:1200 scale

By NEIL RAMAGE

The model was built straight out of the box, being glued rather than clipped together and painted with Tamiya XF19 Sky Grey and XF26 Deep Green. The Swordfish flying overhead are from the Aoshima Royal Navy Plane set in 700 scale giving a forced perspective effect. They were heavily modified with scratch-built wheels, struts, cockpits, propellers, and spinners.

The seascape is built up from layers of Liquitex acrylic gel medium applied over the blue coloured base and painted green when dry. Once the model was set on the base, more acrylic medium was applied along the hull sides and painted white when dry.

# FLYHAWK *HERMES* 1937 1:700 scale

By CHRIS HEWITT

**Chris Hewitt** writes:

HMS *Hermes* was built right out of the box in her pre-WW2 colours. The Swordfish are all white as they were in the late 1930s.

# ORANGE HOBBY *VICTORIOUS* 1:700 scale By RAY NOLTON

**Ray Nolton** writes:

Orange Hobby's 1:700 multi-media kit consists of a superb one-piece waterline casting for the hull, numerous subsidiary resin parts, decals, and a very extensive sheet of fine photo-etch. There are also brass barrels for the 3in/50 and some really nicely produced aircraft. It is not a kit for the faint hearted or beginner: the instructions are relatively clear and well-illustrated but do require some interpretation combined with knowledge and experience of multi-media kits.

Construction is fairly straightforward if complicated. Locating the various resin and etched parts in the box is probably the most time-consuming operation, and I found it very useful to mark off each part on the instructions with a highlighter in order not to lose track of construction. The lattice mast is a rather wonderful piece of brass origami, and very satisfying to complete! Paints were from the Colourcoats and Humbrol ranges.

There is one problem with the kit. It depicts the ship as recommissioned in 1958 but the air group is from after her 1963 refit which altered the ship's appearance. Rather than modifying the ship, I chose to change the aircraft. I obtained the Scimitars and Sea Venoms from Shapeways, and the Skyraiders are from White Ensign. The deck vehicles were modified from US pattern vehicles, and I scratch-built the deck crane which vaguely matches the prototype.

# AOSHIMA *ARK ROYAL* 1:700 scale

By CHRIS HEWITT

The kit was assembled out of the box, the only modification being to remove the catwalks. Most of the detail parts were replaced with aftermarket products. The antenna masts, catwalks, cranes, and mainmast are photo-etch parts from White Ensign; the complex stern catwalk went on very easily. The aircraft and searchlights are from Flyhawk, the boats from Micro Master. The barrels for the 4.5in guns are brass, and there is wood decking under each mounting. The pompoms were from Fine Molds with railing from the White Ensign set applied to each mounting. The finishing touches were rigging from Uschi and figures from Eduard.

The seascape is made from thin layers of wood glue applied over a base painted with two coats of Humbrol 86 Light Olive. Once the model was fixed to the base, gaps were filled with more wood glue, and then white highlights applied with dry brushing before it was sealed with a coat of clear varnish.

## ORANGE HOBBY *VICTORIOUS* 1:700 scale

By CHRIS HEWITT

**Chris Hewitt** writes:

The Orange Hobby kit of the modernised *Victorious* came out in 2014, setting new standards for resin kits with superb detail and plenty of photo-etch parts. There is a small issue as the kit represents her 1958 appearance, but the included air group is from after her 1963 refit when her appearance was altered. Therefore, I decided to modify the ship to her post-1963 appearance by removing two of the 3in/50 mountings and replacing them with extra fuel tanks and accommodation. I also added the 'Alaska Highway' outboard of the island. The decals for the flight deck and aircraft came from the Orange Hobby *Ark Royal* kit released in 2018, which meant the model took four years to complete.

# SCRATCH-BUILT *ILLUSTRIOUS* 1:192 scale

By NORMAN OUGH

This remarkable model was commissioned in June 1949 and completed in April 1951 by master modeller Norman Ough, when the ship was still in active service with all of her late war and immediate post-war modifications in place.

The hull and superstructure are built from wood with the plating details made from paper. The cranes, radars, and masts are made entirely from metal. All the various gun barrels were turned on a lathe from brass wire and aluminum rod. Needless to say, there are no aftermarket items, everything being produced by hand. As Mr Ough put it, 'Any and every means are legitimate in making a model, as long as the effect of reality is achieved.'

The model was for display at the Royal United Service Museum, but it is now in the custody of the National Maritime Museum, Greenwich. When first built it was equipped with naval aircraft reflecting the ship's time as a trials carrier – Attacker, Sea Hornet, Firefly, Sea Fury, and even a Swordfish. At some point the air group was altered to the ten biplanes (presumably meant to be Swordfish) seen here, and the model is now on a base with the 'Battle' class destroyer, HMS *Barfleur*.

A detailed build article is in the 31 May 1951 issue of *The Model Engineer*.

(L2422-003)

*(D7185-1)*

*(L2422-002)*

*(D7185-3)*

*(All photographs © National Maritime Museum, Greenwich, London)*

(L2422

(D7185-7)

(D718

## HMS *Hermes*, China Station 1934

Note that modern research suggests that ships on the China Station were not painted buff and white as widely believed but grey and white.

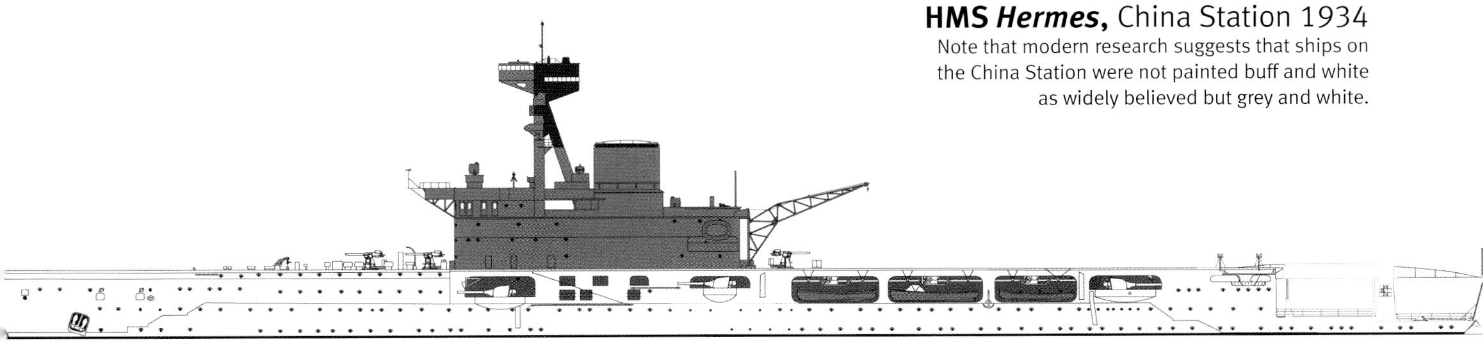

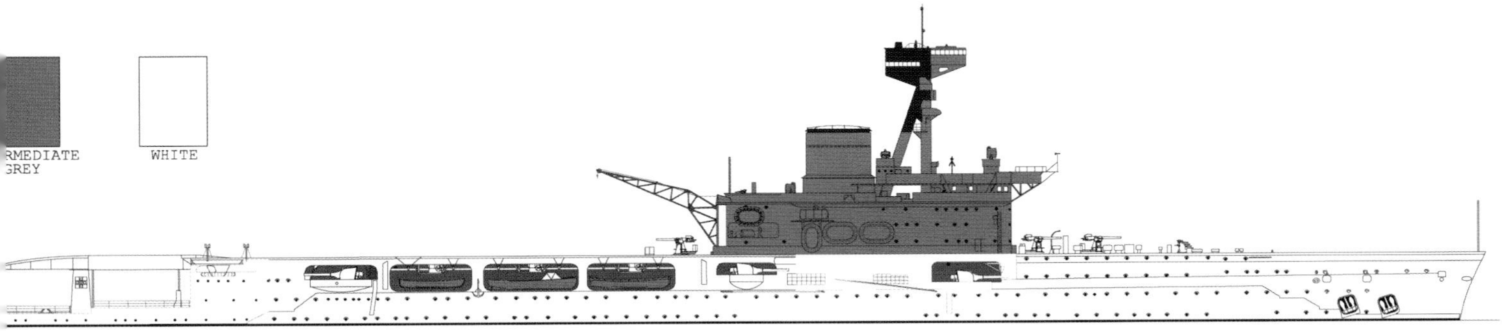


## HMS *Hermes*, 1942

The portside pattern on the island is based on photographic evidence, but the hull patterns remain speculative. It is likely that the panels were the reverse of those on the starboard side; two versions are shown here.

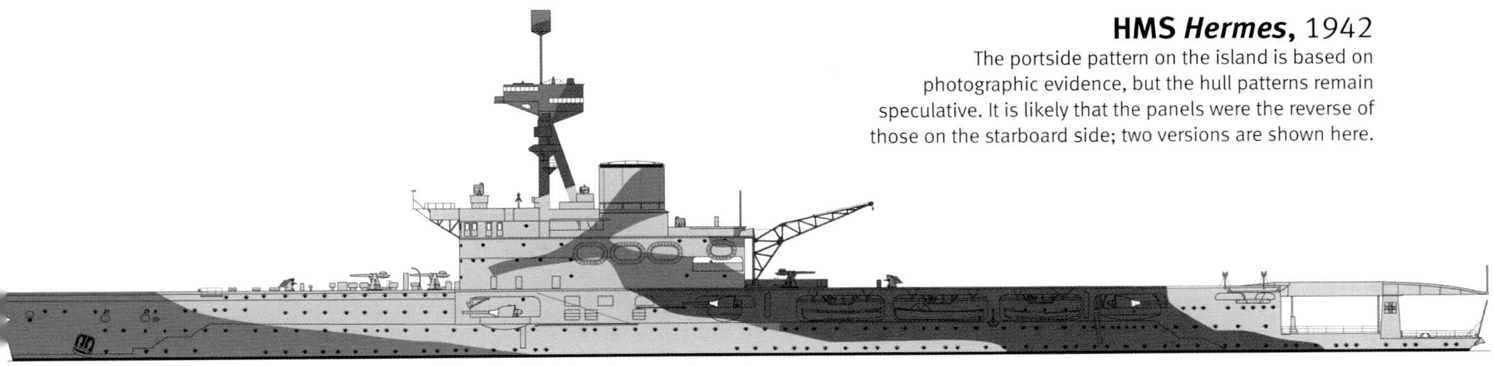

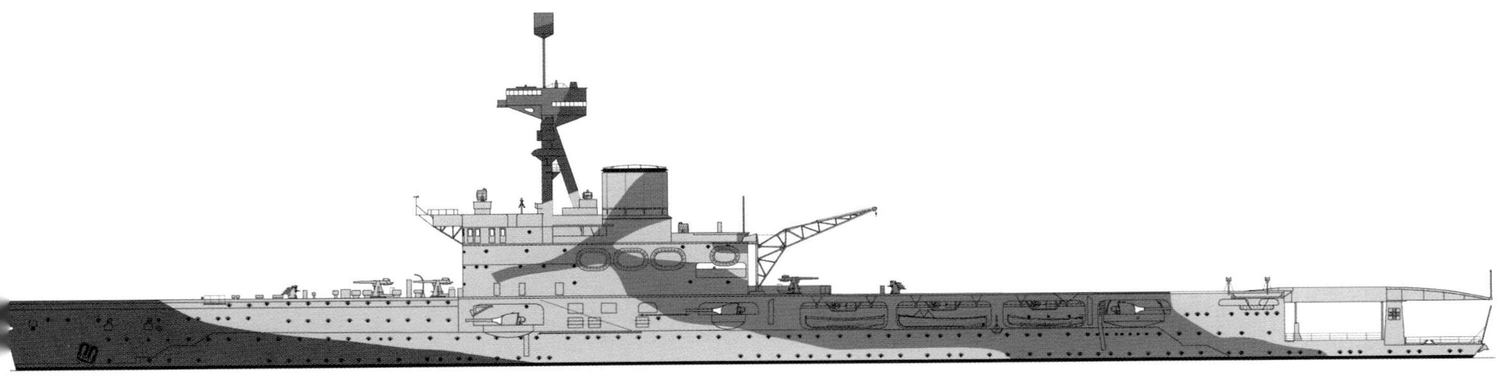

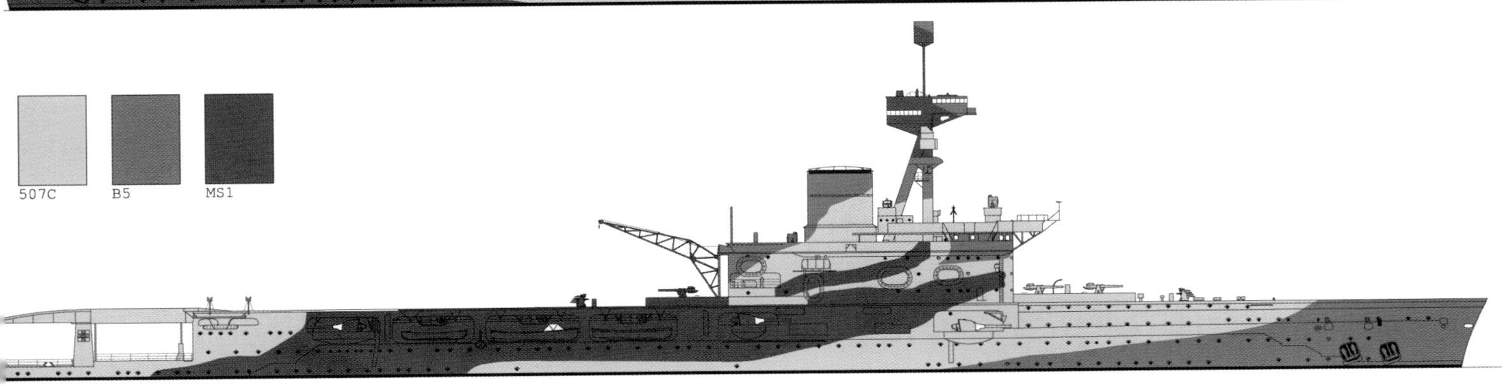

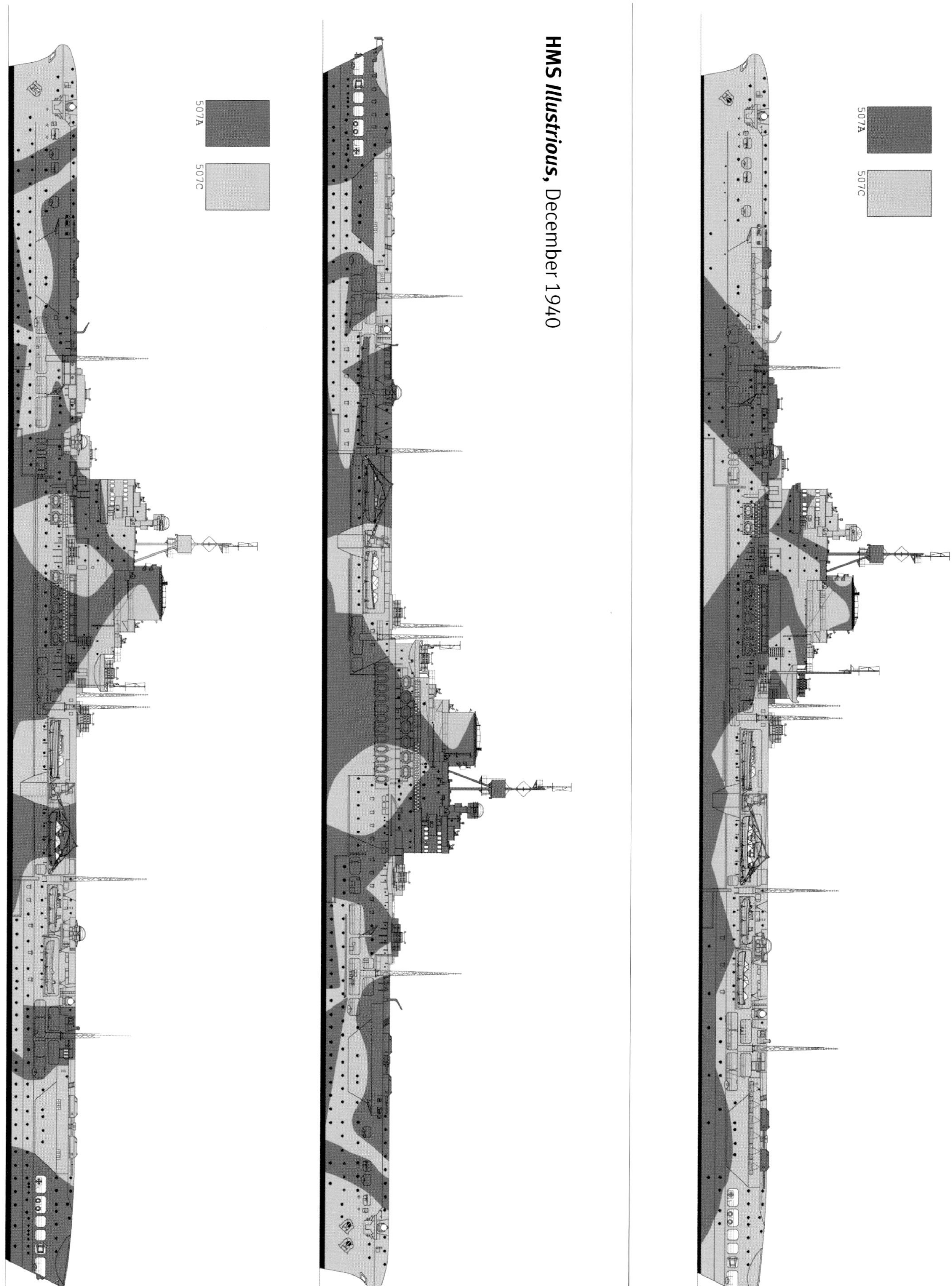


**HMS *Illustrious*,** December 1940

**HMS *Indefatigable*, 1944**

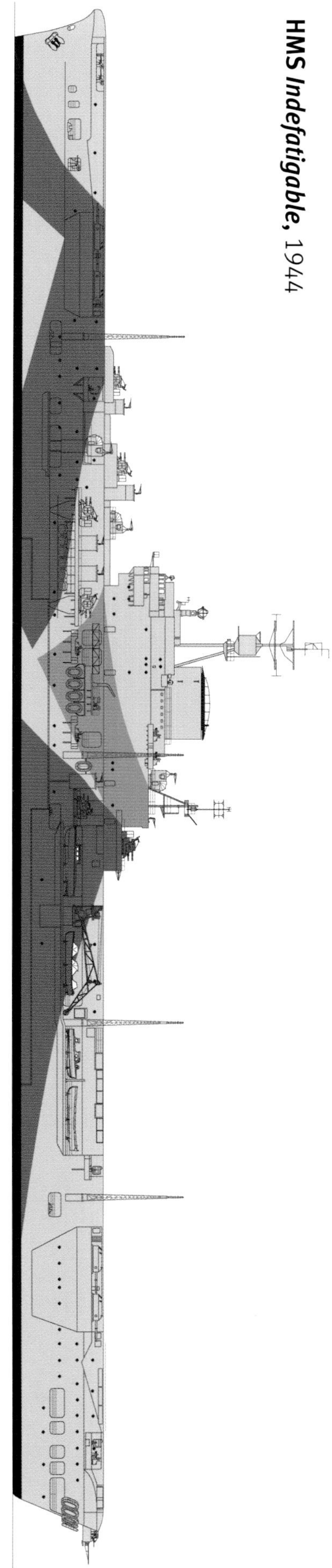

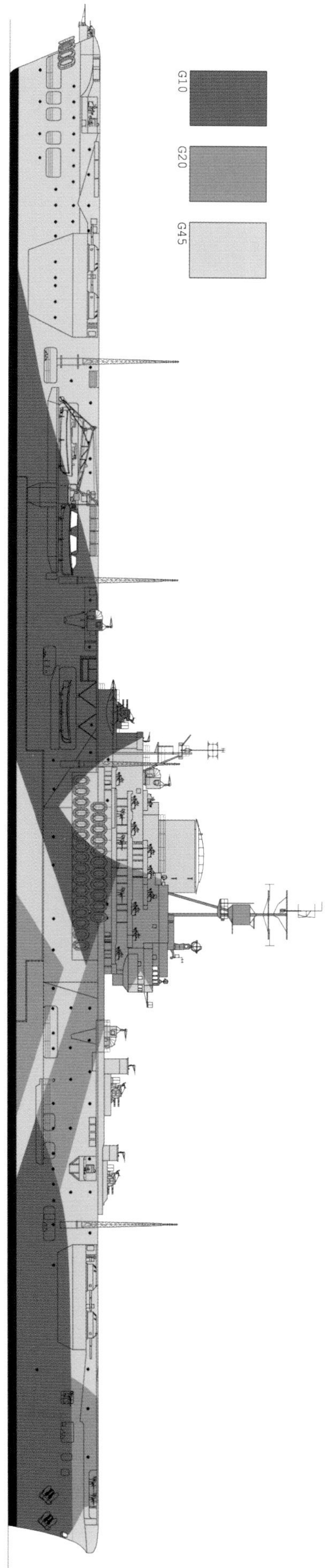


**HMS *Formidable*, 1941**

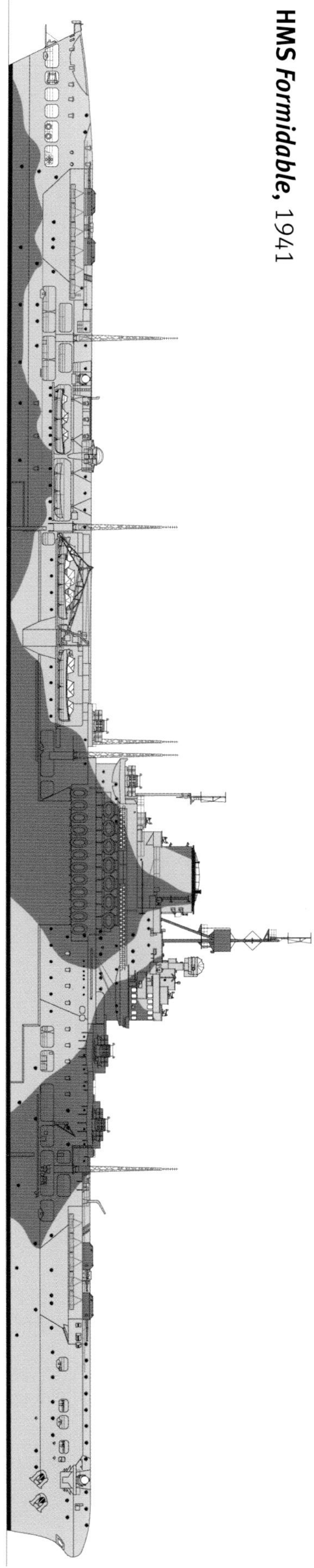

**HMS *Victorious*, 1941**

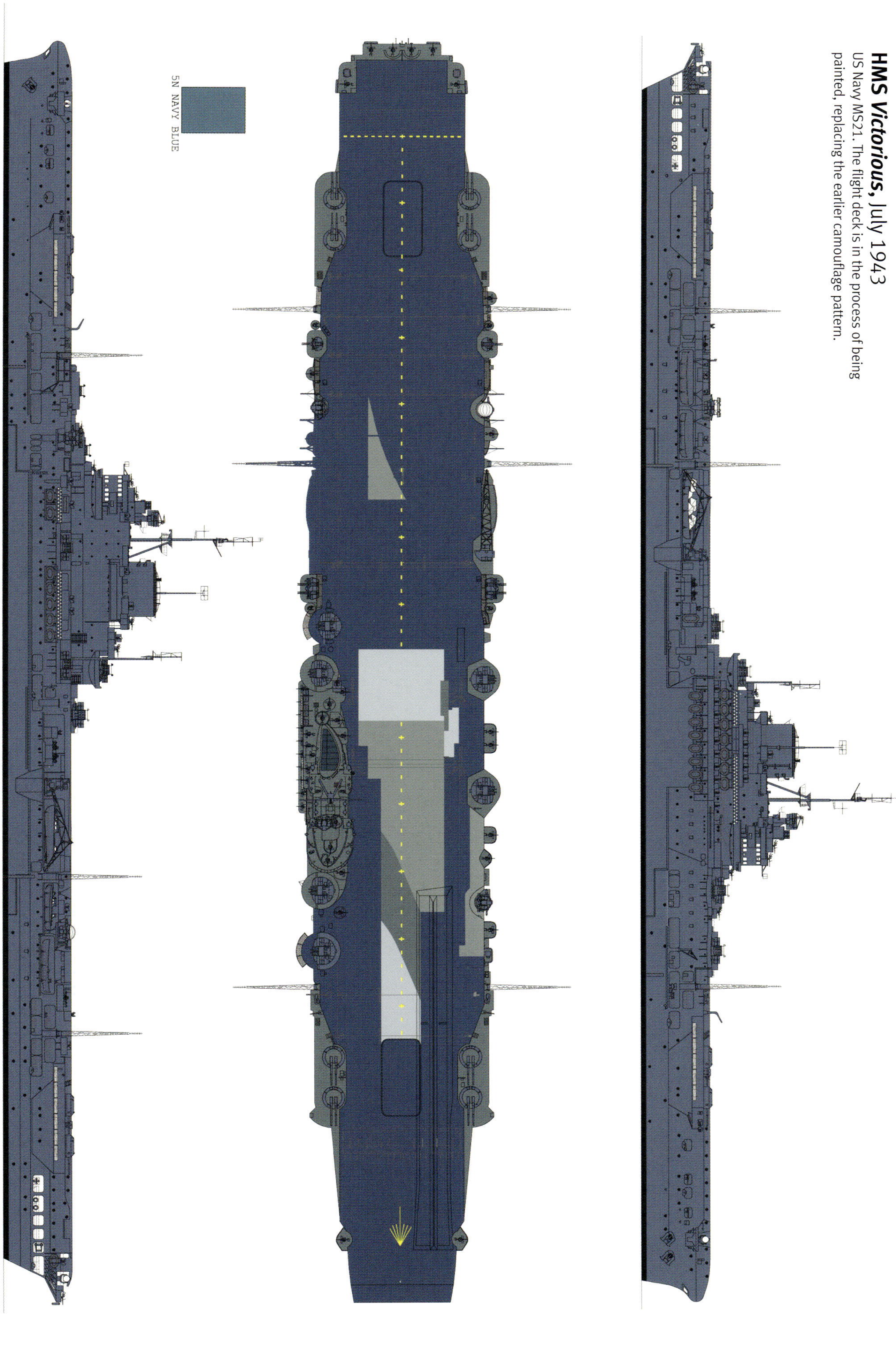


**HMS *Victorious*,** July 1943

US Navy MS21. The flight deck is in the process of being painted, replacing the earlier camouflage pattern.

**HMS *Victorious*, 1945**
Admiralty Standard Scheme

*All drawings by George Richardson, who acknowledges the assistance of Richard Dennis and James Dunn. In the absence of clear and unambiguous information on the paint schemes of these ships, some of the detail is speculative.*

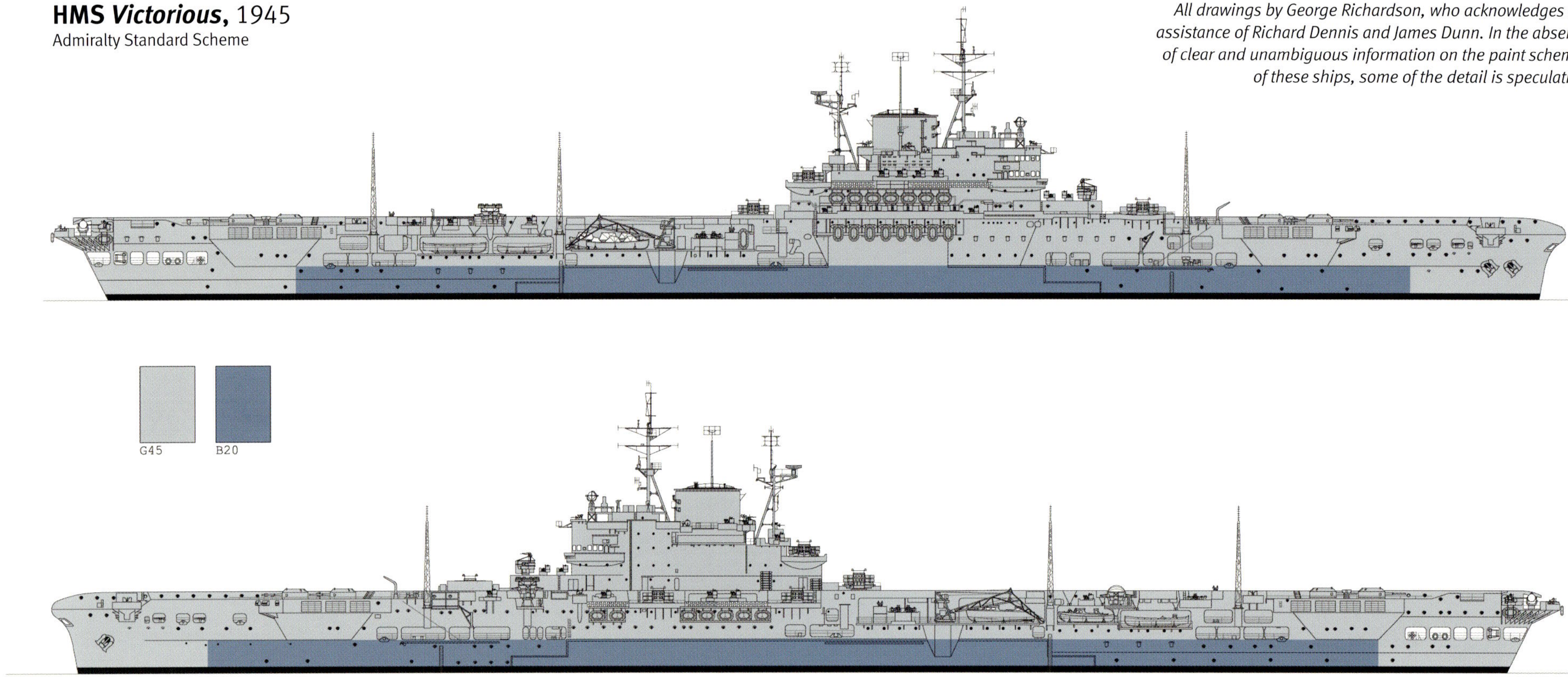


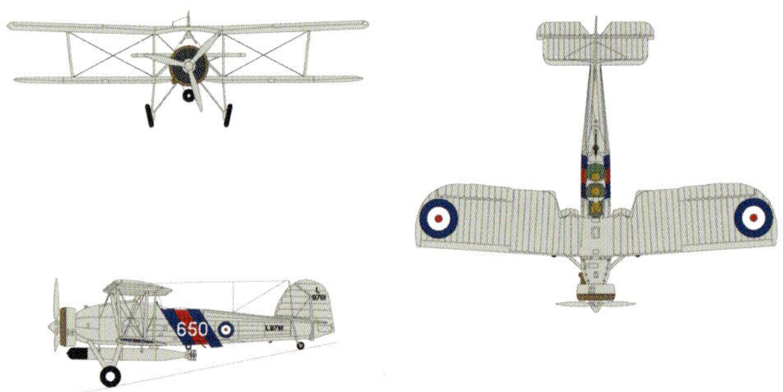

Fairey Swordfish Mk I, pre-war colour scheme

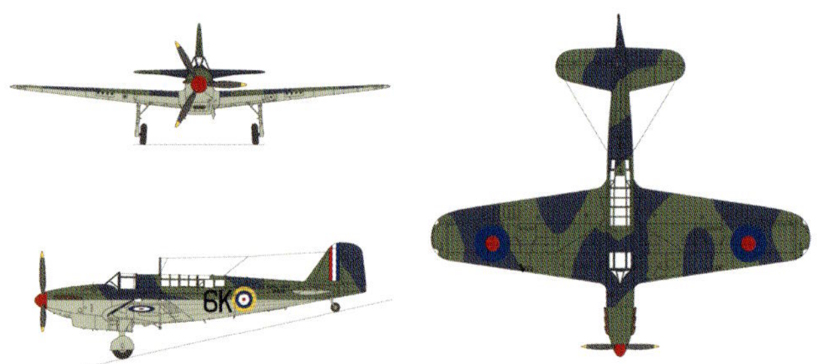

Fairey Fulmar Mk I, 1940

Supermarine Seafire Mk III, British Pacific Fleet 1944

# Alterations

## HERMES

**Refit Date, Alterations**
Commissioned 19.2.1924, **Structure:** Longitudinal arrester wires. **Weapons:** 6 single 5.5in, 4 single 4in

**Malta:** 27.3.1925 to 17.5.1925, **Weapons:** Aftermost 4in gun removed
**Chatham:** 01.11.1927 to 11.1.1928, **Structure:** Longitudinal arrester gear removed. **Weapons:** Two single 2pdr pompoms fitted (one at aft end of island, one aft of after 4in gun in a recessed well)
**Devonport:**.22.9.1933 to 05.10.1934, **Structure:** Transverse arrester gear fitted, spar fitted to control top extending at an angle to port. **Weapons:** Both single 2pdr removed, two 0.5in machine guns fitted (one at aft end of island replacing one 2pdr, one on flight deck forward of 4in mounts)
**Devonport:** 09.1.1940 to 10.2.1940, **Structure:** Fitted with external degaussing cable. **Weapons:** Island-mounted 0.5in machine gun moved aft of after 4in gun
**Simonstown:** 17.8.1940 to 02.11.1940, **Structure:** Collision damage repairs, fitted with internal degaussing cable
**Simonstown:** 05.1.1941 to 20.1.1941, **Structure:** splinter protection fitted to flight deck outboard of 4in guns
**Simonstown:** 18.11.1941 to 31.1.1942, **Structure:** Admiralty Disruptive Camouflage scheme applied
**Electronics:** Type 72 homing beacon fitted to topmast

**Paintwork/Camouflage:**
**Scheme, Dates, Type, Colours**
**1,** 12.1924 to .5.1925, Overall, 507C (Light Grey)
**2,** 6.1925 to .5.1935, RN China Station, White, Intermediate Grey
**3,** 5.1935 to .5.1937, RN China Station, 507C (Light Grey)
**4,** 5.1937 to .8.1939, Overall, 507B, (Dark Grey)
**5,** 9.1939 to .1.1942, Overall, 507C (Light Grey)
**6,** 2.1942 to .4.1942, Disruptive, MS1, 507C, B5 (Dark Grey, Light Grey, Dark Blue)

The ship was assigned to the China Station for five commissions between 1925 and 1937 with refits at home between commissions. The official China Station colour scheme was White Hull/507B Dark Grey superstructure. When wearing this scheme, *Hermes* appears to have

*Hermes* dressed overall at the July 1924 Fleet Review. The back of the mast is painted black. The crew is lining the side and the Royal Marine contingent is drawn up. *(Author's collection)*

Hermes in her final configuration wearing her Disruptive Camouflage Scheme. The Type 72 homing beacon is in place along with splinter shielding for the 4in guns. The island-mounted 0.5in machine gun has been moved to the flight deck aft of the 4in gun. *(Seaforth collection)*

been painted with a lighter Intermediate Grey in place of 507B. The China Station colours were changed to overall 507C in .5.1935.

**Notes:**

The ship was fitted with an external tub for a light machine gun at the front of the control top on the port side, at some time between .11.1934 and .5.1937, and carried it until her loss. There is anecdotal evidence that the ship was fitted with 'a couple' of single 20mm Oerlikons in .3.1942.

## ARK ROYAL

**Refit Date, Alterations**

Commissioned 16.11.1938, **Structure:** 8 arrester wires, 1 safety barrier, 2 catapults. **Weapons:** 8 twin 4.5in, 4 octuple 2pdr pompoms, 8 quad 0.5in. **Electronics:** Type 72X homing beacon

**Liverpool:** 12.1938, **Structure:** Funnel raised by eight feet prior to flying trials, Port side pompom platforms lowered

**Portsmouth:** .4.1939, **Structure:** Lift safety stanchions and protecting ramps removed

**Portsmouth:** 16.2.1940 to 15.3.1940, **Structure:** External degaussing cable fitted

**Liverpool:** 10.10.1940 to 29.10.1940, **Structure:** New simplified flight deck safety barrier installed, Splinter protection fitted to starboard pompoms – zarebas around S1, S2, and S4; S3 fitted with a large metal plate inboard

**Gibraltar:** 21.4.1941 to 24.4.1941, **Weapons:** Two octuple 2pdr pompoms fitted to port side galleries

**Gibraltar:** 30.5.1941 to 04.6.1941, **Structure:** Platform for 3m rangefinder at front of bridge plated in up to the level of the bridge. Windbreak carried around front of new extension; supporting pedestal, charthouse, and rangefinder removed

**Paintwork/Camouflage:**

**Scheme, Dates, Type, Colours,**

**1,** 11.1938 to .11.1941, Overall, 507A (Dark Grey)

The prominent band around the lower portion of the hull seen in some photos is most likely new paint from an interrupted repainting of the entire ship in 1941, and not a camouflage pattern. The island and some portions of the upper hull have the same tone indicating that those areas had also been freshly repainted. The non-repainted portions of the upper hull were in a very sorry state, having been rubbed down/wire brushed and primed in anticipation of finishing the repainting effort. The repainted lower hull band is particularly prominent in photographs taken of the ship as she lay sinking.

**Notes:**

The ship completed with safety stanchions around the lift platforms with ramps to protect them when folded down. For flying trials in 1938 the ship embarked Ospreys and Swordfish; in .9.1939 the air group consisted of Skuas and Swordfish. The first Fulmars were embarked in .10.1940, the ship last carried Skuas in .4.1941. She briefly embarked Albacores from .10.16 to

*Ark Royal* newly completed at Portsmouth in 1939. Seen in this photo are the covers over the pompom directors, the supports for the forward pompom director tub, the rangefinder platform forward of the bridge, the bridge wind deflector, and details of the Carley floats. There is a motor cutter slung outboard on its overhead davits. *(© National Maritime Museum, Greenwich, London N10553)*

19.10.1941 when ferrying them to Malta. The pompom director for S4 meant to be fitted in the port side tub on the rear of the funnel was never installed.

## *ILLUSTRIOUS* CLASS: GENERAL NOTES

The forward arrester wires fitted to *Victorious*, *Formidable*, and *Indomitable* were rarely rigged and were removed in wartime refits. Specific locations for the light AA mounts are given where known. On *Illustrious*, *Victorious*, *Formidable*, and *Indomitable*, there was a large bevel-edged lower casing fitted around the raised catapult. On *Formidable* aft of the lift, this casing was at the same height as the catapult. On *Implacable* and *Indefatigable* the catapult was fitted within a much larger raised casing than on the other ships of the class. This also had a large bevel-edged lower casing fitted around it. The forward windscreen on all six ships was made up of multiple abutting panels except for the two panels on the port side which were separated by the catapult. This arrangement left a slot in the windscreen when it was raised.

## *ILLUSTRIOUS*

**Visual Features:** Two hull openings immediately forward and below the boat galleries on either side. Wind deflector on bridge front not fitted until .11.1941. Flight deck mainmast fitted aft of S3 pompom. Not fitted with tripod mainmast during World War II. HADT 'A' fitted above bridge throughout the war. Type 272 radome on bridge front from .6.1943. Type 72DM homing beacon fitted until .10.1944. Type YE homing beacon fitted in .10.1944.

**Refit Date, Alterations**
Commissioned: 25.5.1940, **Structure:** 9 arrester wires, 2 safety barriers, single catapult. **Weapons:** 8 twin 4.5in, 6 octuple 2pdr pompoms. **Electronics:** Type 72DM homing beacon, Type 79Z radar at tops of foremast (receiving) and mainmast (transmitting)

**Clydebank:** 5.1940, **Structure:** Fitting of external degaussing cable

**Clydebank:** 6.1940, **Structure:** Internal degaussing cable replaced external cable

**Alexandria:** 12.1940, **Structure:** Camouflage Scheme 1 applied

**Alexandria:** 25.1.1941 to 10.3.1941, **Structure:** Temporary repair of bomb damage

**Norfolk, Va:** 12.5.1941 to 28.11.1941, **Structure:** Repair of bomb damage with plans and parts shipped from the UK, useable flight deck length increased to 670ft by building up part of the after round down, new after lift installed, sponsons for AA guns fitted on island on deck 'C', catapult modified to allow tail down launches, catapult upgraded to 20,000lbs, centre propeller shaft was repaired but remained subject to severe vibration at high speeds, bridge wind deflector fitted. **Weapons:** 10 single 20mm added. **Electronics:** Type 285 radar fitted to HADT 'A'

**Liverpool:** 21.12.1942 to .2.1943, **Structure:** Repair of collision damage, camouflage Scheme 2 applied (changed to Scheme 3 by mid-1942). **Electronics:** HF/DF on the foremast removed, Type 281 radar replaced Type 79Z with the same arrangement of aerials; the height discrepancy with the aerials resulted in poor detection ranges

Flight trials in early 1939 with the Hawker Osprey aboard *Ark Royal*. The fuselage chevrons are alternating dark blue, red, and dark blue in colour. The harness attached to the swing-out navigating bridge extends to a support stanchion forward of the funnel. Note the ladder giving access to the pompom directors abaft the funnel and the 'Beware of Airscrews' notice board halfway up the ladder. Pompom director S4 is not fitted in the port side tub. (*Seaforth collection*)

**Liverpool:** 04.3.1943 to 08.6.1943, **Structure:** Two additional arrester wires fitted abaft the after lift, useable flight deck length increased to 695ft by building up part of the remaining after round down and by modifying the bow, flight deck mainmast repositioned to the island abaft the funnel, outriggers fitted for her allocated 10 Seafire IICs, stern AA gallery fitted, submarine lookout positions under the forward 4.5in guns plated over, camouflage Scheme 4 applied. **Weapons:** Forward searchlights replaced with single 20mm, 4 single 20mm fitted in stern gallery, twin 20mm replaced the remaining single 20mm. Light AA modified to: 6 octuple 2pdr pompoms, 10 twin 20mm, 6 single 20mm. **Electronics:** Type 79B radar replaced Type 281 at top of foremast, Type 272 radome fitted to bridge front, Type 285 radar fitted to HADT 'B' and 'X', Mk VI pompom directors with Type 282 radar fitted, VHF Fighter director radio fitted, IFF gear fitted

**Liverpool:** 24.10.1943 to 26.11.1943, **Structure:** Useable flight deck length increased to 740ft by building up the remaining after round down, catapult strengthened to launch fully loaded Barracudas. **Weapons:** Two additional twin 20mm added

**Durban:** 14.8.1944 to 10.10.1944, **Structure:** Alongside refuelling gear fitted, Camouflage Scheme 5 applied. **Weapons:** S2 pompom removed, S4 pompom replaced with single 40mm Bofors. Light AA modified to: 4 octuple 2pdr pompoms, 3 single 40mm, 19 twin 20mm, 14 single 20mm. **Electronics:** Type 72DM homing beacon removed, US 'YE' homing beacon fitted on mast on starboard side of funnel

**Sydney:** 11.2.1945 to 24.2.1945, **Structure:** Centre propeller removed and shaft locked due to deteriorating packing gland and vibration problems; speed limited to 24kts with severe vibration above 19kts

**Rosyth:** 27.6.1945 to .6.1946, **Structure:** Flight deck extended 5ft forward, lifts enlarged to 44ft x 22ft 9in and strengthened, HADT 'A' moved to S2 position, five

The prototype Seafire Mk FIII aboard *Illustrious* in February 1943. The wind deflector has been fitted across the front of the bridge and Type 285 radar is atop HADT 'A'. (*Author's collection*)

bladed prop fitted on the centre shaft, which reduced the vibration, sponsons for single 40mm added to the island, tripod mast fitted to flight deck aft of the island. **Weapons:** 20mm removed except for 12 in the stern galleries, 19 single 40mm fitted. Light AA modified to: 4 octuple 2pdr pompoms, 2 twin 40mm, 19 single 40mm, 12 single 20mm. **Electronics:** Type 79B radar removed, Type 272 radar removed, Type SM-1 radar fitted to bridge roof in place of HADT 'A', Radar Types 293M and 281BQ fitted on YE mast, Type 960 radar fitted on flight deck tripod mast

**Portsmouth:** 1950 to 1951, **Weapons:** Remaining pompoms removed, 20mm removed from stern gallery, 2 20mm removed from port after gallery. Light AA modified to: 2 twin 40mm, 19 single 40mm, 6 20mm. **Electronics:** Modernised Type 277 radar fitted to bridge roof

**Camouflage:**
**Scheme, Dates, Type, Colours,**

**1,** 12.1940 to .2.1942, Disruptive, 507A, 507C (Dark Grey, Light Grey)
**2,** 2.1942 to mid-1942, Disruptive, 507A, 507C, B5 (Dark Grey, Light Grey, Dark Blue)
**3,** Mid-1942 to .6.1943, Disruptive, MS2, MS4A, B5 (Dark Grey, Intermediate Grey, Dark Blue)
**4,** 6.1943 to .10.1944, Disruptive, G5, B15, B30, B55 (Dark Grey, Intermediate Blue, Light Blue, Blue Green)
**5,** 10.1944 to .6.1945, Admiralty Standard, G45, B20 (Light Grey, Dark Blue)

**Notes:**
The ship operated in the Indian Ocean in company with USS *Saratoga* from 27.3. to 18.5.1944. Post-war she served as trials and training carrier until 1954 and was sold for scrapping in 1956.

## *VICTORIOUS*

**Visual Features:** Two hull openings immediately forward and below the boat galleries on either side. Original flight deck mainmast fitted forward of S3 pompom. Mainmast fitted on deck 'C' aft of the funnel in .3.1944. HADT 'A' mounted in place of S2 pompom in .3.1944. Retained her original bow form until modernised in 1950. Quad 40mm fitted to boat galleries in .4.1943. Completed with Type 285 radar on HADT 'A'. Type 272 radar lantern fitted on bridge top in .10.1942. Type 72M homing beacon fitted until January 1943. Type YE homing beacon fitted in January 1943. Only 6 antenna masts after .4.1943.

**Refit Date, Alterations**
Commissioned: 29.3.1941, **Structure:** 9 arrester wires aft, 2 arrester wires forward, 2 safety barriers, single catapult, camouflage Scheme 1. **Weapons:** 8 twin 4.5in, 6 octuple 2pdr pompoms. **Electronics:** Type 72DM homing beacon, Type 79Z radar at tops of foremast (receiving) and mainmast (transmitting), Type 285 radar fitted to HADT 'A'

Prior to .4.1942, **Electronics:** Type 285 radar fitted to HADT 'B' and 'X', Type 282 radar fitted to five pompom directors, the director in the port side sponson was not fitted, MF/DF aerial fitted to the front of the island. **Weapons:** Three 20mm singles added (one aft of S1 pompom, one on island deck 'A' aft of S2 pompom, one on island deck 'D' on sponson on port side of funnel)

**Rosyth:** 9.1942 to .10.1942, **Structure:** Camouflage Scheme 2 applied. **Electronics:** Type 272 radar in lantern housing fitted on top of the signal bridge, HF/DF loop removed from the foremast

**Norfolk, Va:** 01.1.1943 to 30.1.1943, **Structure:** Useable flight deck length increased to 695ft by building up part of the after round down, stern 20mm gallery fitted, forward arrester wires removed, additional arrester wire fitted aft of the elevator. **Weapons:** 18 single 20mm added (5 in gallery starboard side of island on deck 'B', 1 on starboard side of funnel at deck 'D', 6 in stern gallery, 1 in sponson ahead of HADT 'X', 1 in sponson ahead of HADT 'Y', 1 ahead of HADT 'B', 2 in sponson abreast port boiler room vents, 1 in front of P1 pompom). Light AA modified to: 6 octuple 2pdr pompom, 21 single 20mm. **Electronics:** Type 72DM homing beacon removed, US 'YE' homing beacon fitted on mast on starboard side of funnel

**Pearl Harbor:** 09.3.43 to 08.4.1943, **Structure:** Forward transmitter antenna mast removed from both sides, camouflage Scheme 3 applied. **Weapons:** 1 quad 40mm with Mk 51 directors fitted on boat gallery forward of the crane, port and starboard, 1 twin 40mm with Mk51 directors forward on sponsons replaced after searchlights, port and starboard, 10 single 20mm added in flight deck gallery on starboard side of island, 4 single 20mm replaced the pompom directors on the island, 2 single 20mm fitted between P1 and P2 pompoms, 2 single 20mm replaced forward searchlights, 2 single 20mm removed from stern gallery, replaced with repositioned searchlights. Light AA modified to: 6 octuple 2pdr pompom, 2 quad 40mm, 2 twin 40mm, 37 single 20mm

**Liverpool:** 03.1.1944 to 04.3.1944, **Struc-**

**ture:** Useable flight deck length increased to 740ft by building up the remaining after round down, two extensions fitted on island deck 'D' above the Admiral's viewing platforms on deck 'C' for additional air lookout positions, HADT 'A' moved to S2 position, telescopic RDF mast on flight deck removed, new mainmast fitted aft of the funnel, support braces angled forward, four pompom directors reinstalled (one forward atop the bridge, one to starboard of the new mainmast, one between the island and forward pompom on the flight deck, one in a port side sponson), starboard side hull galleries aft of boat galleries plated over, submarine lookout positions under the forward 4.5in guns plated over, Camouflage Scheme 4 applied. **Weapons:** S2 pompom removed, 4 twin 20mm replaced 5 single 20mm on deck 'B' platform, 2 twin 20mm replaced 2 outboard single 20mm on stern platform, 4 twin 20mm replaced 4 single 20mm on port side, 1 twin 20mm replaced 1 single 20mm at each end of flight deck gallery on starboard side of the island, 1 twin 20mm replaced 1 single 20mm on either side of funnel on deck 'D', 1 twin 20mm replaced 1 single 20mm forward of HADT 'X', 1 twin 20mm replaced 1 single 20mm forward of HADT 'Y', 1 twin 20mm replaced 1 single 20mm on deck 'B' gallery, 4 single 20mm replaced by pompom directors, 1 single 20mm forward of island removed. Light AA modified to: 5 octuple 2pdr pompom, 2 quad 40mm, 2 twin 40mm, 17 twin 20mm, 14 single 20mm. **Electronics:** Type 279 radar replaced Type 79, Type 277 replaced Type 272 on signal bridge, TBS aerials, US 'SG' radar with Type 252 IFF, and a 'hayrake' IFF interrogator for the Type 279 fitted to foremast, Type 281B radar, Type 293 radar, Type 242 IFF interrogator, and Type 243 IFF interrogator fitted to new mainmast

**Bombay:** 30.9.1944 to 05.10.1944, **Structure:** Fitted with new rudder

**Bombay:** 16.11.1944 to 22.11.1944, **Structure:** Steering gear repairs

**Sydney:** 2.1945, **Weapons:** 2 single 40mm Boffin replaced 2 twin 20mm on flight deck gallery starboard side of the island, 2 single 40mm replaced 2 single 20mm in forward searchlight sponsons, 2 single 40mm replaced outboard twin 20mm and searchlights on stern platform. Light AA modified to: 5 octuple 2pdr pompom, 2 quad 40mm, 2 twin 40mm, 4 single 40mm, 2 single 40mm Boffin, 13 twin 20mm, 12 single 20mm

**Plymouth:** 10.1945 to .12.1945, **Structure:** Accommodations for trooping in the hangar

**Plymouth:** 25.6.1947 to .9.1947, **Structure:** More accommodations and classrooms added

**Camouflage:**

**Scheme, Dates, Type, Colours**

**1,** 5.1941 to .10.1942, Disruptive, 507A, MS3, B6, White (Dark Grey, Intermediate Grey, Light Blue, White)

**2,** 10.1942 to .3.1943, Disruptive, Unknown,

**3,** 3.1943 to .3.1944, US Navy, 5N Navy Blue (Dark Blue)

**4,** 3.1944 to .10.1945, Admiralty Standard, G45, B20 (Light Grey, Dark Blue)

**Notes:**

*Victorious* was 5ft longer than her sisters *Illustrious* and *Formidable*. At Pearl Harbor in .3.1943, it was found that the arrester wires were too thick to engage the hooks of the Avengers which now formed part of her air group and the arrester wire tensioning was not sufficient to trap aircraft landing off the centre line; none of the shipboard cranes including the mobile deck crane could lift an Avenger. The arrester gear and the Avenger tail hooks were modified. After post-war trooping voyages, she was part of the Portland training squadron from 1947 to 1950 before being taken in hand for modernisation in 1950; she subsequently served until 1967.

*Victorious* in March 1943 showing the 20mm AA gallery, the three rows of her Signal Group lights for night flying, and liferafts. *(Author's collection)*

## FORMIDABLE

**Visual Features:** Two hull openings immediately forward and below the boat galleries on either side. Original flight deck mainmast fitted forward of S3 pompom. Tripod mainmast fitted to flight deck immediately aft of the island after .3.1944. HADT 'A' mounted in place of S2 pompom in .3.1944. Retained her original bow form until scrapped. Type 72M homing beacon fitted until .3.1944. Type YE homing beacon fitted in .3.1944.

**Refit Date, Alterations**
Commissioned: 24.11.1940, **Structure:** 9 arrester wires aft, 2 arrester wires forward, 2 safety barriers, single catapult. **Weapons:** 8 twin 4.5in, 6 octuple 2pdr pompoms. **Electronics:** Type 72DM homing beacon, Type 279 radar at tops of foremast (receiving) and mainmast (transmitting)

**Norfolk, Va:** 26.8.1941 to 29.11.1941, **Structure:** Battle damage repair, forward arrester wires removed, useable flight deck length increased to 629ft by building up part of the after round down. **Weapons:** 10 single 20mm added

**Belfast:** 21.12.1941 to 03.2.1942, **Structure:** Collision damage repaired, submarine lookout positions under the forward 4.5in guns plated over, stern AA gallery fitted, Camouflage Scheme 2 applied. **Electronics:** Type 279 radar replaced Type 281 on both masts, Type 285 radar fitted to HADT 'A' and 'Y'

**Belfast:** 29.11.1943 to 31.3.1944, **Structure:** Useable flight deck length increased to 740ft by building up the remaining after round down, HADT 'A' moved to S2 position, flight deck mainmast removed, tripod mainmast fitted aft of island, Camouflage Scheme 3 applied. **Weapons:** S2 pompom removed. Light AA modified to: 5 octuple 2pdr pompom, 1 quad 2pdr pompom, 20 twin 20mm, 14 single 20mm. **Electronics:** Type 72DM homing beacon removed, Type 277 radar installed on bridge roof, US 'YE' homing beacon fitted on mast on starboard side of funnel, Type 281 and Type 293 radars fitted to mainmast

**Sydney:** 30.5.1945 to 27.6.1945, **Structure:** Battle damage repairs and upgrades to be operated as a flagship, port side hull gallery aft of the boat galleries plated over. **Weapons:** Light AA modified to: 5 octuple 2pdr pompom, 1 quad 2pdr pompom, 5 single 40mm, 7 40mm Boffin, 11 twin 20mm, 12 single 20mm

**Camouflage:**
**Scheme, Dates, Type, Colours**
**1,** 3.1941 to .2.1942, Disruptive, 507A, 507C (Dark Grey, Light Grey)
**2,** .2.1942 to .3.1944, Disruptive, MS1, MS4A, B5, B6 (Dark Grey, Intermediate Grey, Dark Blue, Light Blue)
**3,** 3.1944 to mid-1944, Disruptive, G5, B15, B30, B55 (Dark Grey, Dark Blue, Light Blue, Blue Green)
**4,** mid-1944 to .12.1945, Admiralty Standard, G45, B20 (Light Grey, Dark Blue)
**5,** 12.1945 to .2.1946, Dark Hull, Light Upper Works, G10, G45 (Dark Grey, Light Grey)

**Notes:**
*Formidable*'s last day as an operational carrier was 24.8.1945; she never operated aircraft after that. Used for trooping after the war, she went into reserve in .8.1947 and never returned to active service. She was proposed for modernisation but was found unsuitable due to unrepaired action damage and poor upkeep; sold for scrap in Jan 1953, the first of the class to be disposed of.

Folded Fulmars on *Formidable*'s flight deck. Note the deck-edge details: crane, antenna masts and arrester gear. (*Seaforth collection*)

## INDOMITABLE

**Visual Features:** Nine small rectangular hull openings forward of the island on the starboard side. Amidships boat stowage is on two levels. The deck extension ahead of the forward 4.5in galleries is at a lower level than the rest of the class. Supporting sponsons for the after 4.5 galleries are very shallow. Enlarged forward lift. HADT 'A' moved to S2 position in .10.1950. Retained her original bow form until modernised in 1950. Quad 40mm fitted to boat galleries in .4.1944. 20mm galleries fitted to quarterdeck hull openings in .4.1944. Completed with Type 285 radar on HADT 'A'. Type 271 radar lantern on bridge front from .2.1943 to .4.1944. Type 72DM homing beacon fitted until .4.1944. Type YE homing beacon fitted in .4.1944. Type 277Q radar dishes at each end of the island fitted in .10.1950.

**Refit Date, Alterations**
Commissioned: 10.10.1941, **Structure:** 9 arrester wires aft, 3 arrester wires forward, 3 safety barriers, single catapult, Camouflage

*Indomitable* after her 1950 refit showing her new radar suite with two Type 277 on the island roof, Type 293Q atop the foremast, and Type 960 on the mast aft of the funnel. The YE aerial has been transferred to the stub mast at the front of the funnel. Five 40mm mounts can be seen on the island and the ship retains three of her pompoms. HADT 'A' has been moved to the flight deck in front of the island. Sea Hornets and Sea Furies are at the front of the flight deck. (*By courtesy of Nigel Denchfield*)

Scheme 1. **Weapons:** 8 twin 4.5in, 6 octuple 2pdr pompoms, 8 single 20mm. **Electronics:** Type 72DM homing beacon, Type 79Z radar at tops of foremast (receiving) and mainmast (transmitting), Type 282 radar fitted to pompom directors, Type 285 radar fitted to HADT 'A'

**Norfolk, Va:** 10.11.1941 to 21.11.1941, **Structure:** Repairs to damage from grounding 3.11.1941 at Kingston, Jamaica

**Liverpool:** 25.8.1942 to 11.2.1943, **Structure:** Repairs to action damage, forward arrester wires removed, submarine lookout positions under the forward 4.5in guns plated over. **Weapons:** 8 twin 20mm replaced 8 single 20mm, 2 single 20mm fitted in forward searchlight positions. Light AA modified to: 6 octuple 2pdr pompoms, 8 twin 20mm, 2 single 20mm. **Electronics:** Type 79B radar replaced Type 79Z on mainmast, Type 281B radar replaced Type 79Z on foremast, Type 271 radar lantern fitted on front of island

**Norfolk, Va:** 31.8.1943 to 15.4.1944, **Structure:** Repairs to action damage, additional AA sponsons fitted including stern gallery, Camouflage Scheme 3 applied. **Weapons:** 2 quad Bofors fitted to boat gallery immediately forward of the crane, one on each side, 2 twin Bofors fitted to sponsons just forward of after 4.5in gallery, one on each side, 2 twin 20mm fitted to port side sponson between port side pompoms, 2 twin 20mm fitted to deck 'C' sponson on starboard side of island, 4 twin 20mm fitted to flight deck level sponson on starboard side of island, 4 twin 20mm fitted to sponsons on after corners of flight deck, two on each side, 6 single 20mm fitted to sponsons below quarterdeck hull openings, three on each side, 4 single 20mm fitted to forward searchlight positions, two on each side, 6 single 20mm fitted to stern gallery. Light AA modified to: 6 octuple 2pdr pompoms, 2 quad 40mm, 2 twin 40mm, 20 twin 20mm, 16 single 20mm. **Electronics:** Type 72DM homing beacon removed, Type 271 radar removed, US 'YE' homing beacon fitted on mast on starboard side of funnel, Type SM-1 radar fitted atop the front of the bridge, Type 285 radar fitted to HADT 'X'

**Sydney:** 06.6.1945 to 18.7.1945, **Structure:** Collision damage repairs after colliding with HMS *Quilliam*. **Weapons:** Light AA modified to: 6 octuple 2pdr pompoms, 2 quad 40mm, 2 twin 40mm, 12 single 40mm, 1 40mm Boffin, 12 twin 20mm, 12 single 20mm

**Portsmouth:** 7.1950 to .10.1950, **Structure:** Limited modernisation, bow rebuilt, deck extension ahead of forward 4.5in galleries plated over, HADT 'A' moved to S2 position, catapult, arrester wires, and barriers upgraded to handle 20,000lbs. **Weapons:** S2 pompom removed. **Electronics:** Type 960 radar fitted aft of the funnel, Type 293Q radar fitted atop the foremast, Type 963 radar fitted on after part of the island, two Type 277Q fitted fore and aft atop the island

**Camouflage:**
**Scheme, Dates, Type, Colours,**

**1,** 10.1941 to .6.1942, Disruptive, 507A, MS4A, B5, B6, White (Dark Grey, Intermediate Grey, Dark Blue, Light Blue, White)

**2,** 6.1942 to .4.1944, Disruptive, MS2, MS4A, B5 (Dark Grey, Intermediate Grey, Dark Blue)

**3,** 4.1944 to .11.1944, Disruptive, G5, B15, B30, B55 (Dark Grey, Dark Blue, Light Blue, Blue Green)

**4,** 11.1944 to .3.1946, Admiralty Standard, G45, B20 (Light Grey, Dark Blue)

**Notes:**
The ship completed without the after round down. Corsairs could only be accommodated in her short lower hangar, which was given over to accommodation space during Pacific operations in 1945. For this reason she operated Hellcats and Avengers exclusively from .11.1944 onwards. After post-war trooping voyages, she went into reserve until 1950 when she once again became fully operational after a limited modernisation. In .6.1953 she hosted deck landing trials of a Sea Vampire using the prototype mirror landing system, after which she went into reserve, being sold for scrap in .9.1955.

## *IMPLACABLE*

**Visual Features:** Starboard side: one hull opening aft and two forward of the boat gallery, with all three below the level of the boat gallery. Port side: both large hull openings aft of the boat gallery are below the level of the boat gallery. Port side: large sponson above after hull openings containing AA weapons. Flight deck considerably bulged to port just forward of the after 4.5in gallery. Enlarged forward lift. Port side seaplane crane fitted forward of the starboard side seaplane crane. Very deep sponsons supporting after 4.5in galleries. Conspicuous director towers for S1 and S2 pompoms. Only one octuple pompom aft of the island. Larger island and funnel. Tripod mainmast at rear of island. HADT 'A' on fight deck forward of island. HADT 'Y' on island aft of funnel. Type 72 Homing Beacon still fitted at end of war. Type YE homing beacon fitted by 1952.

**Refit Date, Alterations**
Commissioned: 22.5.1944, **Structure:** 8 arrester wires, 3 safety barriers, single catapult, Camouflage Scheme 1. **Weapons:** 8 twin 4.5in, 5 octuple 2pdr pompoms, 2 twin 20mm island 'A' deck starboard, 4 twin 20mm island 'B' deck starboard, 4 twin 20mm island 'C' deck starboard, 2 twin 20mm island 'C' deck port, 1 twin 20mm immediately forward of aft flight

*Implacable* passing under Lions Gate bridge inbound to Vancouver on 11 October 1945. Note the flight deck vehicles, S1 and S2 pompoms with their directors, HADT director, the mobile deck crane parked forward of the island, safety barrier stanchions, the ship's DUKW parked aft of the island, and the 40mm on the island. The crew is lining the side while the ship's band is performing, and there are sports pitches chalked on the flight deck. *(By courtesy of David Hobbs)*

A good view of *Implacable*'s island and port side pompoms at Vancouver in October 1945. The ship's crest and battle honours are on the side of the island. (*By courtesy of Vancouver Maritime Museum*)

deck edge starboard, 2 twin 20mm immediately forward of aft flight deck edge port, 2 twin 20mm between P1 and P2 pompoms port, 1 twin 20mm forward of after 4.5in guns port, 2 twin 20mm on deck sponson aft of boat gallery port, 3 single 20mm island 'A' deck starboard, 2 single 20mm immediately aft of forward flight deck edge starboard, 2 single 20mm immediately aft of forward flight deck edge port, 1 single 20mm in upper hangar deck opening aft of anchors starboard, 1 single 20mm in upper hangar deck opening aft of anchors port, 2 single 20mm in aftermost hull opening aft starboard, 2 single 20mm in aftermost hull opening aft port, 4 single 20mm in stern gallery under aft edge of flight deck. Light AA fit: 5 octuple 2pdr pompoms, 20 twin 20mm, 17 single 20mm. **Electronics:** Type 72DM homing beacon, Type 277 radar on bridge roof, Type 281 radar on tripod mainmast, Type 282 radar on pompom directors, Type 285 radar on all HADT, Type 293 radar on foremast above homing beacon

**Rosyth:** 16.6.1944 to 16.8.1944, **Structure:** Engine defects repaired

**Rosyth:** 15.12.1944 to 10.3.1945, **Structure:** Repairs to bow damaged in storm on 28.11.1944, Camouflage changed to Scheme 2. **Weapons:** 3 quad 2pdr pompoms fitted (1 on sponson aft of island starboard side, 1 on port side boat gallery forward of the crane, 1 on port side forward and above boat gallery), all twin 20mm replaced with single 40mm, 1 twin 20mm fitted below S1 pompom starboard, 1 twin 20mm fitted ahead of HADT director port. Light AA modified to: 5 octuple 2pdr pompoms, 3 quad 2pdr pompoms, 20 single 40mm, 2 twin 20mm, 17 single 20mm

**Sydney:** 15.3.1946 to 29.4.1946, **Structure:** Camouflage changed to Scheme 3. **Weapons:** Light AA modified to: 5 octuple 2pdr pompoms, 3 quad 2pdr pompoms, 8 single 40mm, 4 single 40mm Boffin, 8 twin 20mm, 14 single 20mm

**Camouflage:**
**Scheme, Dates, Type, Colours**

**1,** 5.1944 to .3.1945, Disruptive, G10, G45, B30 (Dark Grey, Light Grey, Blue)

**2,** 3.1945 to .4.1946, Admiralty Standard, G45, B20 (Light Grey, Dark Blue)

**3,** 4.1946 to 1947, Dark Hull Light Upper Works, G10, G45 (Dark Grey, Light Grey)

**Notes:**
*Implacable* operated an air group of 48 Seafires, 12 Fireflies, and 21 Avengers in the Pacific; even with 94,650gal of avgas there was only enough for 5 sorties per aircraft. She returned Canadian POWs to Vancouver in .10.1945.

Post-war modernisation was planned, stripping the ship down to the lower hangar deck and building up a new 17ft 6in hangar with a new deck above for accommodations, an armoured flight deck, 2 large centreline lifts, a new island, and 3in/50 AA mounts. These plans were cancelled due to the escalating costs of the modernisation of *Victorious*. Any other use of the ship post-war was limited by the low height of her hangars. She could not even be used as a commando carrier as the Westland Whirlwind stood 15ft 4in high.

From .10.1947, she was Home Fleet deck training carrier and carried out operational trials of the Gloster Meteor in .6.1948. In .9.1949, she embarked Sea Vampires. She remained operational until 1950 when she became a seagoing training ship without aircraft until 1954 when she was paid off for scrapping; being dismantled from .11.1955.

## INDEFATIGABLE

**Visual Features:** Starboard side: one large hull opening forward of the boat gallery, at the same level as the boat gallery; one small hull opening forward and one aft of the boat gallery, both below the level of the boat gallery. Port side: both large hull openings aft of the boat gallery were at the same level as the boat gallery; another small hull opening is immediately aft. Flight deck considerably bulged to port just forward of the after 4.5in gallery. Enlarged forward lift. Port side seaplane crane fitted forward of the starboard side seaplane crane. Very deep sponsons supporting after 4.5in galleries. Conspicuous director towers for S1 and S2 pompoms. Only one octuple pompom aft of the island. Larger island and funnel. Tripod mainmast at rear of island, initially fitted directly aft of funnel. HADT 'A' on fight deck forward of island. HADT 'Y' on island aft of funnel was not fitted until .11.1944. Type 72 Homing Beacon still fitted at end of war. Type 72 Homing Beacon removed by 1950. Never fitted with YE homing beacon.

**Refit Date, Alterations**
Commissioned: 03.5.1944, **Structure:** 8 arrester wires, 3 safety barriers, Single catapult, Camouflage Scheme 1. **Weapons:** 8 twin 4.5in, 5 octuple 2pdr pompoms, 2 twin 20mm island 'A' deck starboard, 4 twin 20mm island 'B' deck starboard, 4 twin 20mm island 'C' deck starboard, 2 twin 20mm island 'C' deck port, 1 twin 20mm immediately forward of aft flight deck edge starboard, 2 twin 20mm immediately forward of aft flight deck edge port, 2 twin 20mm between P1 and P2 pompoms port, 1 twin 20mm forward of after 4.5in guns port, 3 single 20mm island 'A' deck starboard, 2 single 20mm immediately aft of forward flight deck edge starboard, 2 single 20mm immediately aft of forward flight deck edge port, 1 single 20mm in upper hangar deck opening aft of anchors starboard, 1 single 20mm in upper hangar deck opening aft of anchors port, 2 single 20mm in aftermost hull opening aft starboard, 2 single 20mm in aftermost hull opening aft port, 4 single 20mm in stern gallery under aft edge of flight deck. Light AA fit: 5 octuple pompoms, 18 twin 20mm, 17 single 20mm. **Electronics:** Type 72DM homing beacon, Type 277 radar on bridge roof, Type 281 radar on tripod mainmast, Type 282 radar on pompom directors, Type 285 radar on all HADT, Type 293 radar fitted to foremast above homing beacon

**Clydebank:** 28.9.1944 to 08.11.1944, **Structure:** HADT 'Y' fitted to after end of island, mainmast moved to aft end of island. **Weapons:** 1 quad 2pdr pompom fitted port side boat gallery forward of the crane, 1 single 40mm fitted below S1 pompom starboard, 1 single 40mm fitted ahead of HADT director port. Light AA modified to: 5 octuple 2pdr pompoms, 1 quad 2pdr pompom, 2 single 40mm, 18 twin 20mm, 17 single 20mm
**Leyte:** 20.4.1945 to 01.5.1945, **Structure:** Repairs from kamikaze strike on 1.4.1945
**Sydney:** 05.6.1945 to 07.7.1945, **Structure:** Camouflage changed to Scheme 3. **Weapons:** Light AA modified to: 5 octuple 2pdr pompoms, 1 quad 2pdr pompom, 5 single 40mm, 5 single 40mm Boffin, 14 twin 20mm, 12 single 20mm

**Camouflage:**
**Scheme, Dates, Type, Colours**
**1,** 12.1943 to .2.1945, Disruptive, G10, G20, G45 (Dark Grey, Intermediate Grey, Light Grey)
**2,** 3.1945 to .7.1945, Admiralty Standard, G45, B20 (Light Grey, Dark Blue)
**3,** 7.1945 to .4.1946, Dark Hull Light Upper Works, G10, G45 (Dark Grey, Light Grey)

**Notes:**
*Indefatigable* carried out deck landing trials of the Mosquito bomber on 15.3.1944. During the Okinawa campaign she operated an air group of 40 Seafires, 20 Avengers, and 9 Fireflies. The ship was present at the formal surrender ceremonies in Tokyo Bay on 2.9.1945. Reduced to reserve at Portsmouth on 7.1.1947 after post-war trooping service, she was reactivated as a training ship from 1950 to 1954 with no aircraft carried. Decommissioned in .10.1954, she was sold for scrap in .9.1956.

She was operational as a carrier from .12.1943 to .3.1946, just 27 months.

## *VICTORIOUS* MODERNISED

**Refit Date, Alterations**
Commissioned: 14.1.1958, **Structure:** 4 arrester wires, 1 safety barrier, 2 steam catapults. **Weapons:** 6 twin 3in/50, 1 sextuple 40mm Bofors. **Electronics:** Type 957 Tacan navigation beacon, Type 984 radar, Type 293Q radar, Type 963 radar, Type 978 radar

**Portsmouth:** 26.2.1960 to 15.8.1960, **Structure:** Bridle catcher fitted ahead of port catapult. **Electronics:** VHF/DF antenna replaced by plate shaped UHF/DF antenna
**Portsmouth:** 02.4.1962 to 12.6.1963, **Structure:** Bridle catcher removed, Deck Landing Projector Sights replaced the Deck Mirror Landing Sights; the new units used a projected rather than a reflected light, Flyco altered to an angular shape, platform fitted outboard of the island for a mobile equipment servicing station, known as the 'Alaska Highway' to the crew, boat stowage modified to include mobile trollies. **Weapons:** 2 3in/50 mounts removed (middle mounting on each side), sextuple Bofors removed. **Electronics:** Port side antenna masts replaced with 4 whip aerials

**Notes:**
The newly modernised ship served in the Atlantic, in the Mediterranean, and in the Far East, participating in many exercises with NATO and other friendly aligned countries. She was in the Far East during the Indonesian Confrontation as part of British forces defending the newly created country of Malaysia.

On 11.11.1967, as *Victorious* was nearing the end of a scheduled refit in Portsmouth, a fire destroyed a crew's mess area, one rating losing his life. While damage was minimal, the government of the day used this as an excuse to withdraw her from service as part of a scheme to eliminate the RN carrier forces. She was sold for scrap in .7.1969, far short of her expected hull life. One of her former captains, Vice Admiral Sir Richard Janvrin gave her a final salute from the last serviceable Swordfish as she left Portsmouth under tow on 11.7.1969.

## HMS *Hermes*, 1934

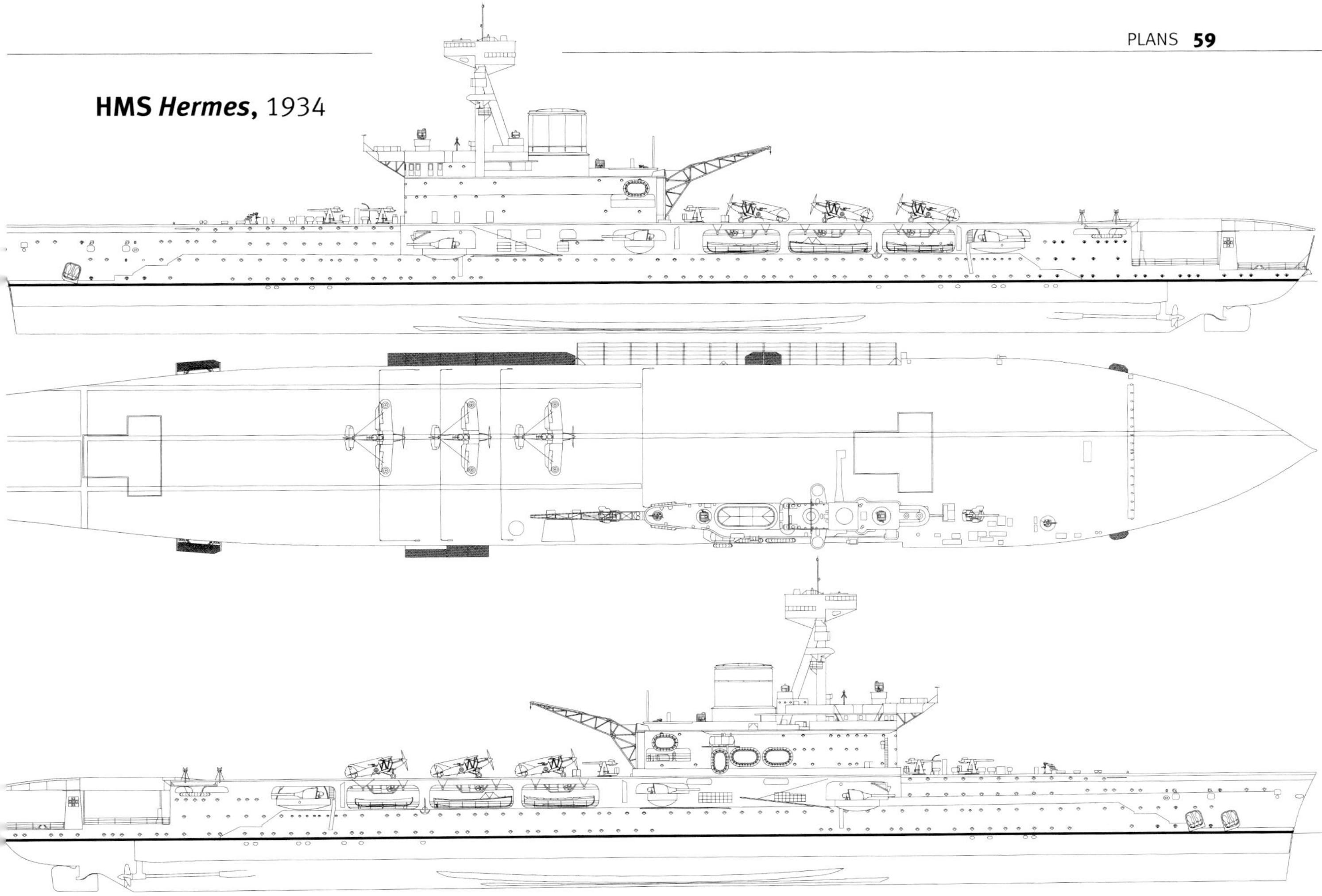

## HMS *Hermes*, 1942

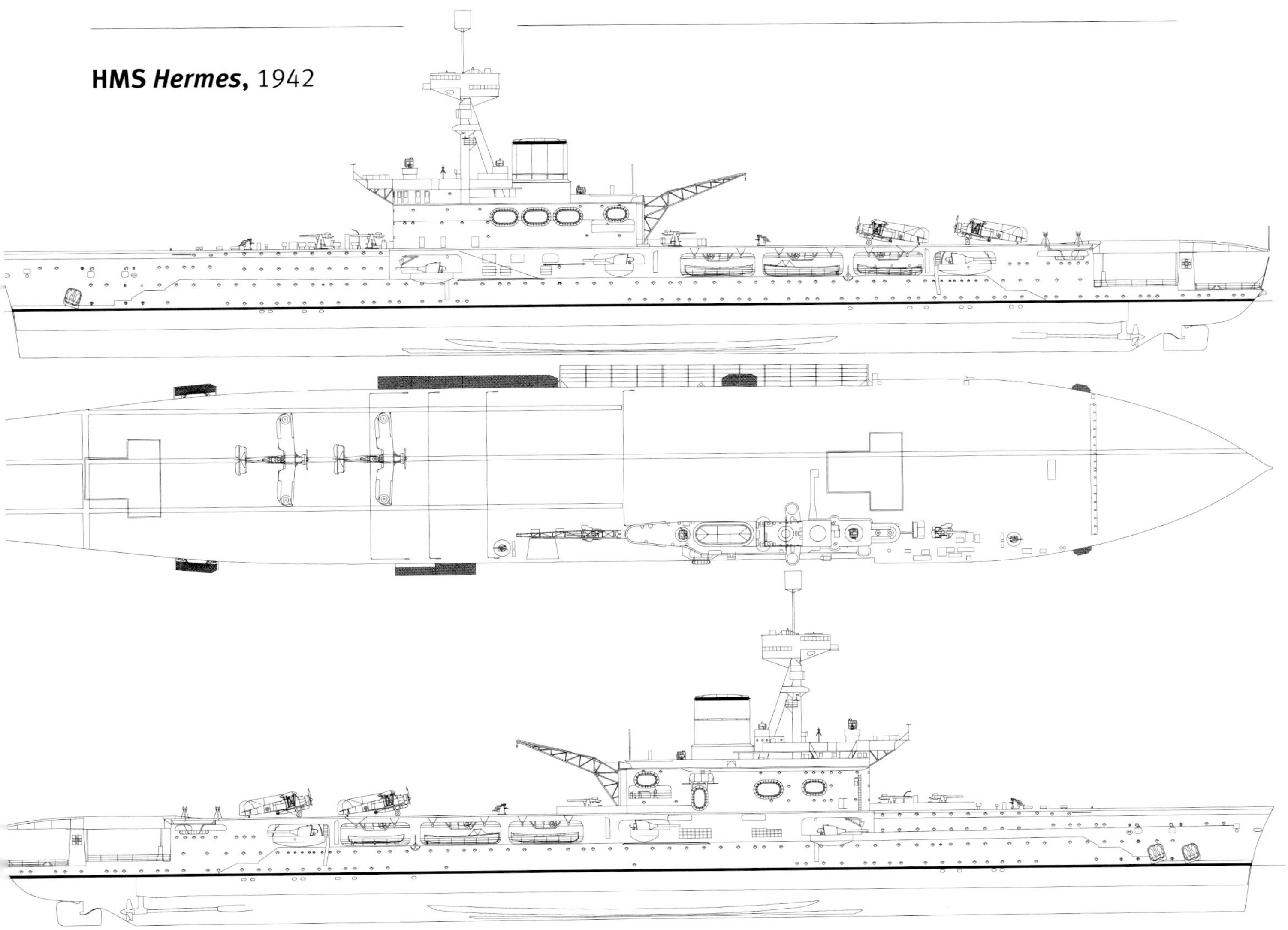

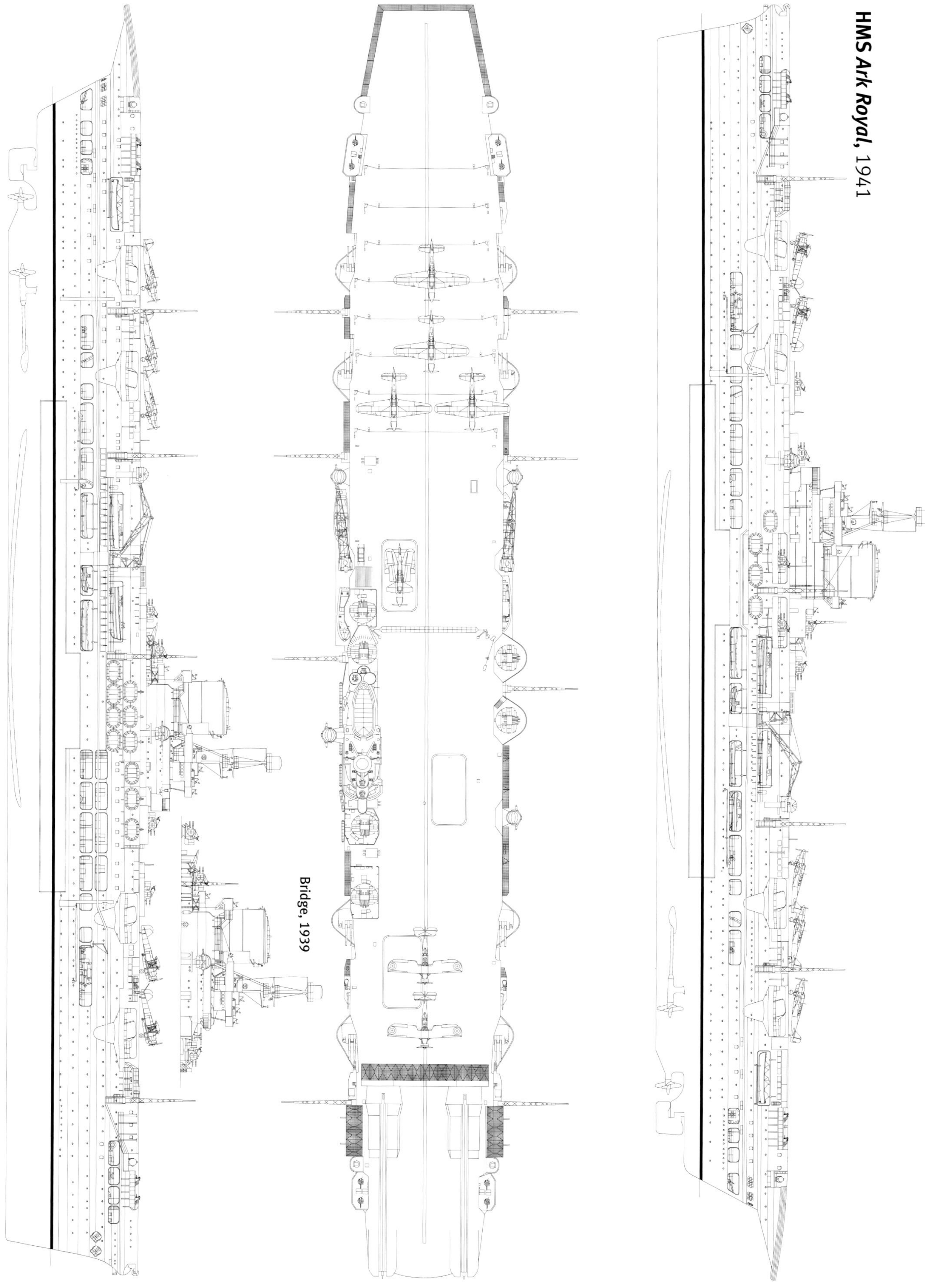


**HMS *Ark Royal*, 1941**

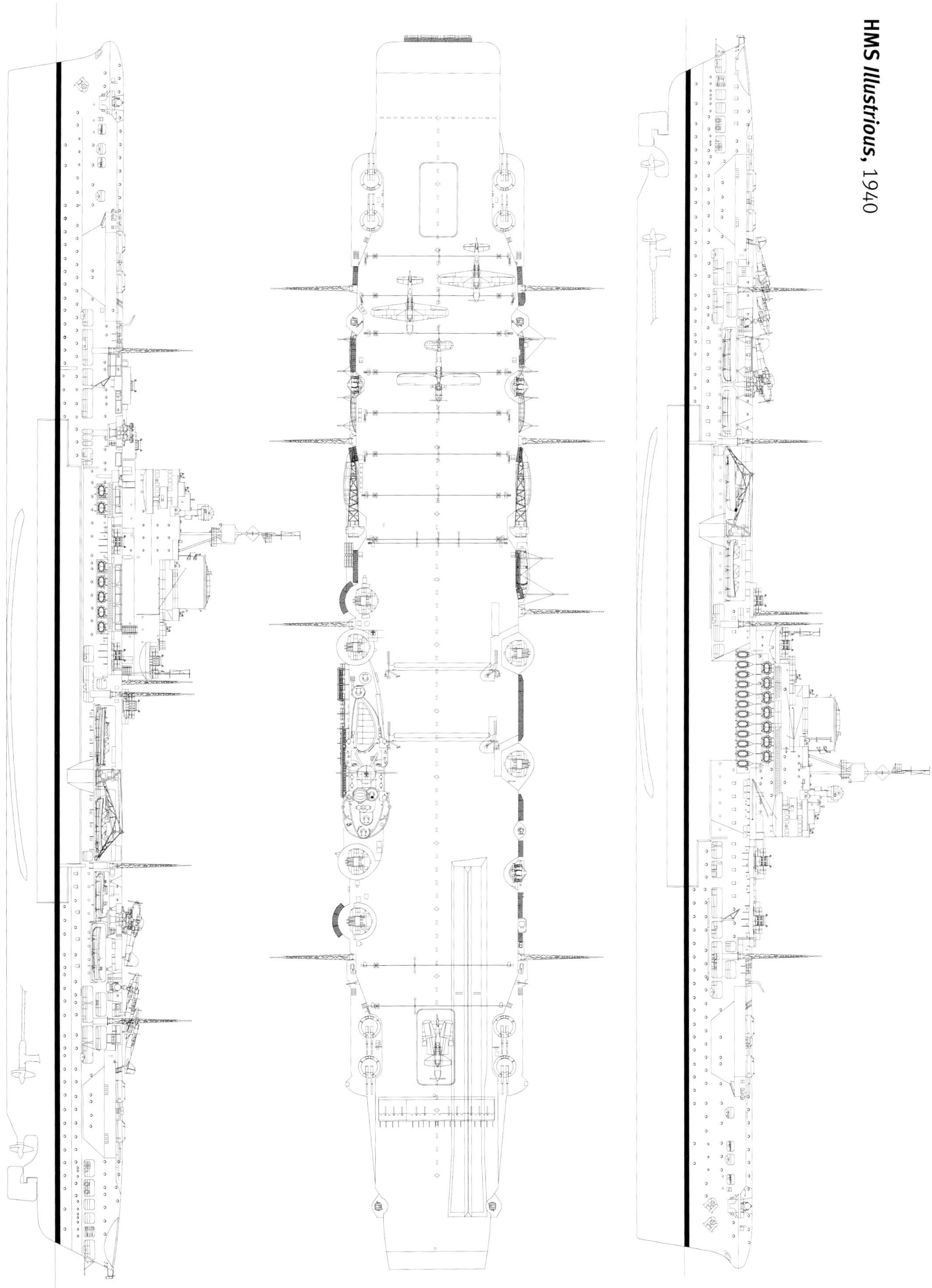

HMS *Illustrious*, 1940

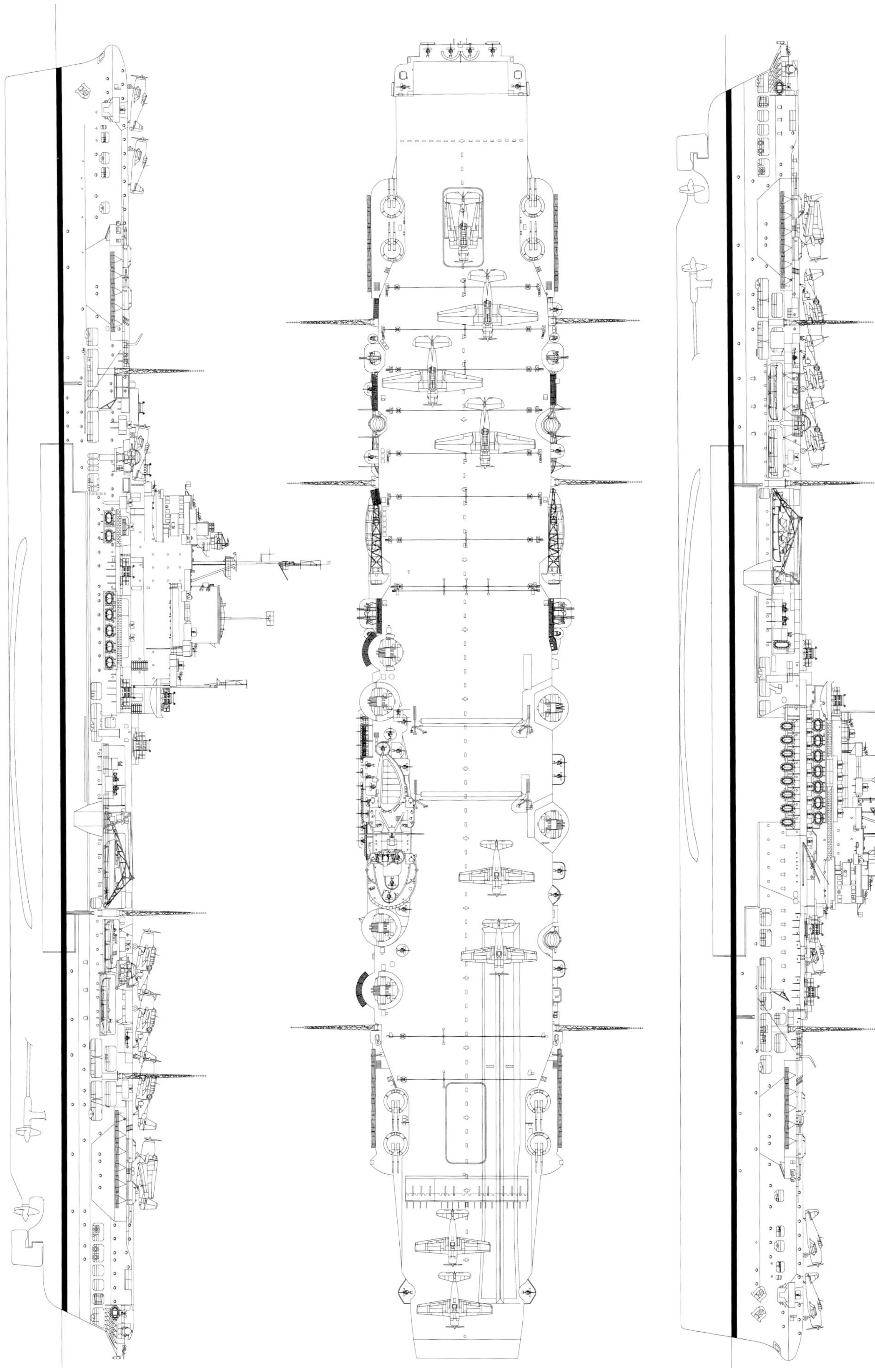

**HMS *Victorious*,** 1943

## HMS *Indefatigable*, 1944

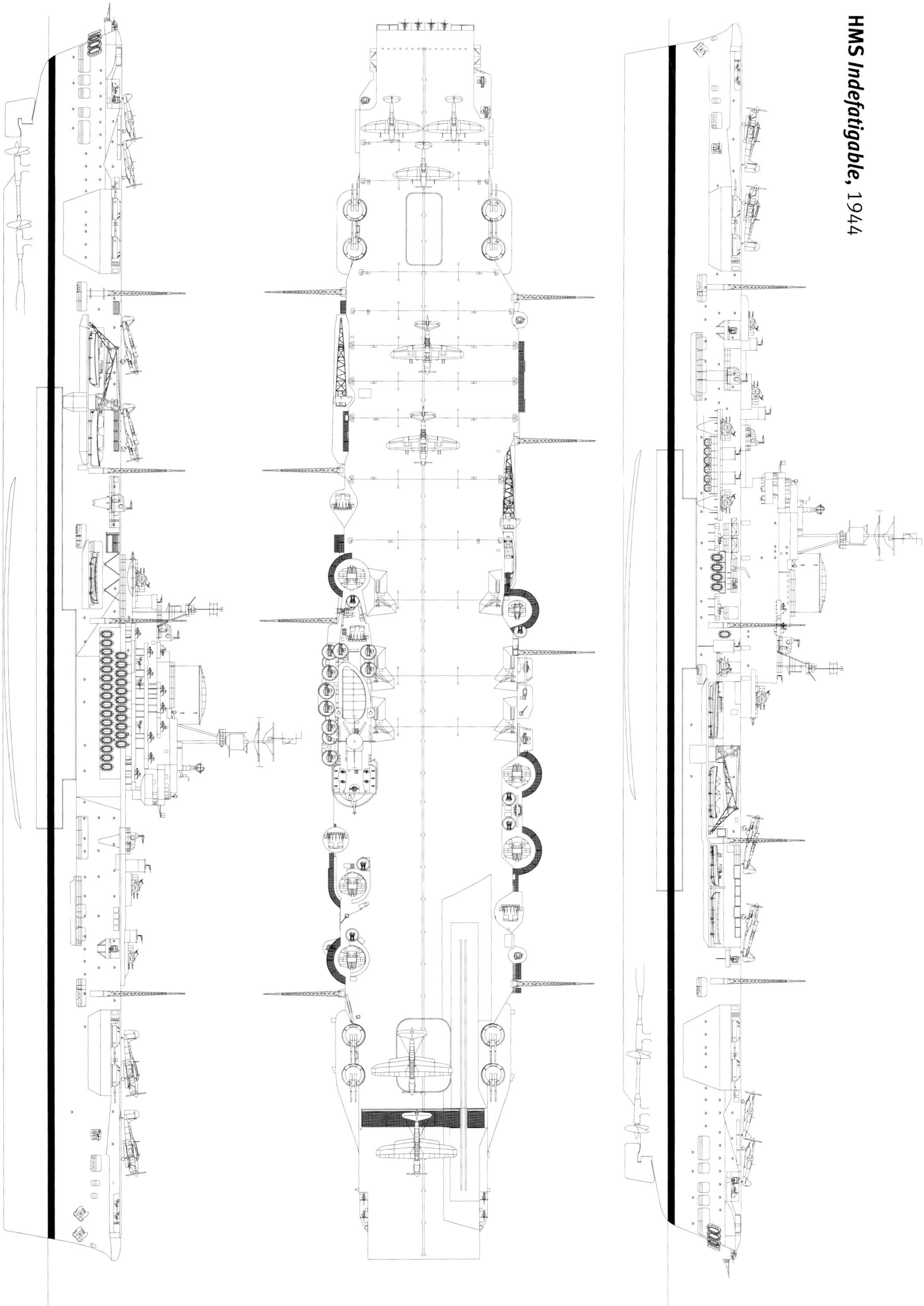

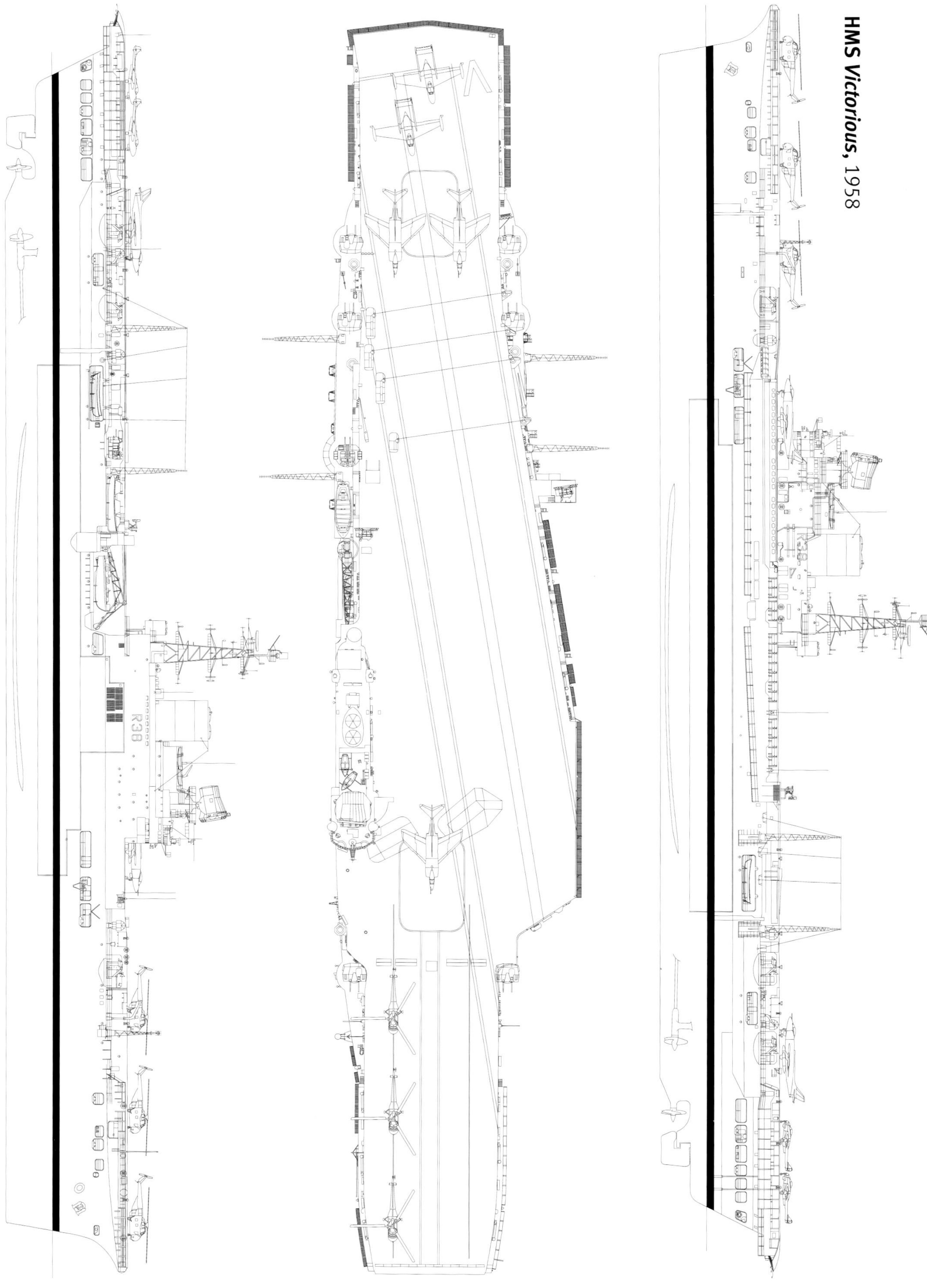


**HMS *Victorious*, 1958**